Canned Citrus Processing

Canned Citrus Processing

Techniques, Equipment, and Food Safety

Yang Shan

Editor-in-Chief

AMSTERDAM • BOSTON • HEIDELBERG • LONDON
NEW YORK • OXFORD • PARIS • SAN DIEGO
SAN FRANCISCO • SINGAPORE • SYDNEY • TOKYO
Academic Press is an imprint of Elsevier

Academic Press is an imprint of Elsevier
125, London Wall, London EC2Y 5AS, UK
525 B Street, Suite 1800, San Diego, CA 92101-4495, USA
225 Wyman Street, Waltham, MA 02451, USA
The Boulevard, Langford Lane, Kidlington, Oxford OX5 1GB, UK

ISBN: 978-0-12-804701-9

British Library Cataloguing-in-Publication Data
A catalogue record for this book is available from the British Library

Library of Congress Cataloging-in-Publication Data
A catalog record for this book is available from the Library of Congress

Publisher: Nikki Levy
Acquisition Editor: Simon Tian
Editorial Project Manager: Naomi Robertson
Production Project Manager: Melissa Read
Designer: Matthew Limbert

Typeset by TNQ Books and Journals
www.tnq.co.in

Printed and bound in the United States of America

Contents

Preface vii

1 Overview of the Canned Citrus Industry 1
1.1 Industry Background 1
1.2 History of Development 3
1.3 Current Industrial Situation 5

2 Canned Citrus Processing 7
2.1 Background 8
2.2 Categories of Canned Foods 10
2.3 Preservation Principles of Canned Foods 11
2.4 Procedures and Techniques Used for Manufacturing Canned Citrus Products 18
2.5 Common Quality Issues and Controlling of Canned Foods 36
2.6 New Sterilization Techniques of Canned Foods 38

3 Machinery and Equipment for Canned Citrus Product Processing 47
3.1 Material Handling Machinery and Equipment 48
3.2 Cleaning Machinery and Equipment 51
3.3 Machinery and Equipment for Processing Raw Citrus Materials and Semifinished Products 59
3.4 Exhaust and Sterilization Machinery and Equipment 76
3.5 Packaging Machinery and Equipment 86
3.6 Typical Canned Citrus–Processing Production Line 104

4 Quality and Safety Control during Citrus Processing 105
4.1 Limits and Requirements for Pesticide Residues, Contaminants, and Additives in Citrus and Canned Products According to National Standards 106
4.2 Hazard Analysis and Critical Control Points for Canned Citrus Processing 106
4.3 GMP Control during the Processing of Canned Citrus Products 123
4.4 Construction of the Traceability Management System for Canned Citrus Products 143

Appendix A: Traceability Management Forms 151
Appendix B: Examples of RFID Traceability Identification Tags 159

Appendix C: Informative Appendix **161**
Appendix D: Standard Catalog of Canned Citrus and Codex Stan 254-2007 .... **163**
References **173**

Index **177**

Preface

Citrus is the largest fruit crop in the world. In 2011, the citrus plantation area was 8.73 million hectares and produced a yield of 131 million tons worldwide. Citrus and its products are ranked third in the global agricultural commodity trade. China is the leading producer of citrus in the world. In 2011, China's citrus plantation area was 2.28 million hectares and citrus production was 29.4 million tons. In China, the citrus industry employs 180 million citrus farmers.

Canned citrus is a major product in the citrus processing industry. Canned citrus products account for 80% of the processed citrus in China, and their exports account for 75% of the global market. The Chinese canned citrus industry is very competitive globally. Since the "Ninth 5-year Plan (1996–2000)," Chinese scientists have resolved several issues that transpired with canned citrus products. Scientists have eliminated the odor of dimethyl sulfide caused by high-temperature sterilization, reduced the use of acidic or alkaline chemicals, improved the brittleness of the citrus segments, and decreased the precipitation of hesperidin. Furthermore, Chinese scientists have used ethylene-vinyl alcohol (EVOH) plastic packages instead of fragile glass bottles or stiff metal cans, dealt with waste water generated by enzymatic hydrolysis, and removed the peel and sac coating of citrus fruits. At present, the quality of Chinese canned citrus has gained international recognition; thus, these products are in great demand in the European, American, and Japanese markets. In addition to Spain, China has become the new "Kingdom of Canned Citrus."

The canned citrus industry is one of the most important sectors in food industry. It has a long research history in the food science or food engineering field. However, monographs on the canned citrus industry in China are rare. To reflect the technical and developing trends in the Chinese canned citrus industry, the author has written this book titled, *Canned Citrus Processing: Techniques, Equipment, and Food Safety*. This book is based on a collection of literature and documents regarding research in the citrus industry as well as achievements and experiences at home and abroad.

As the chief editor, Yang Shan edited this book with Gaoyang Li, Jianxin He, Donglin Su, Juhua Zhang, Fuhua Fu, Qun Zhang, Yuehui Wu, Xiangrong Zhu, Wei Liu, Lvhong Huang, Jiajing Guo, Qiutao Xie, and other colleagues who participated in the compilation. It includes four chapters, references, and four appendixes. This book contains an overview of canned citrus industry, canned citrus processing machinery and equipment, canned citrus processing quality and safety control, and corresponding standards at home and abroad. Moreover, this book has the characteristics of complete content, and it is a scientific and practical manual, which will provide the reference of the canned citrus processing for scientists, scholars, and students from advanced institutes, universities, and citrus-processing enterprises, and many farmers from cooperative organizations related to citrus processing.

Any useful suggestions or criticisms regarding the information in this book will be highly appreciated.

Editors
December 5, 2013

Chapter | One

Overview of the Canned Citrus Industry

CHAPTER OUTLINE

1.1 Industry Background 1
1.2 History of Development 3
  1.2.1 Starting Stage (1959–1969) 3
  1.2.2 Development Stage (1970s to the Mid-1980s) 3
  1.2.3 Prosperous Stage (1990s to the Present Day) 3
1.3 Current Industrial Situation 5

Canned citrus is a food with isolated segments that are preserved in a specific container with sugar solution after its sac coating is removed. Because canned citrus products are sterilized after complete sealing, the inherent freshness and nutrition are maintained for a relatively long period. Mandarin oranges, grapefruit, oranges, and hybrid tangerines are frequently used as raw materials for canned citrus products. The canned citrus industry encompasses all activities related to the processing of canned citrus products, which mainly includes citrus plantation, product transportation, packaging or container processing, and processing equipment manufacturing.

1.1 INDUSTRY BACKGROUND

Citrus fruit is harvested in 146 countries including China, Brazil, the USA, and the Mediterranean, and this fruit is predominantly produced worldwide. The Food and Agriculture Organization of the United Nations (FAO) statistics show that in 2011, approximately 8.73 million hectares were used to plant citrus, the citrus yield was 131 million tons worldwide, and 23.2 billion US dollars were spent internationally trading citrus and citrus products. Citrus is the third largest agricultural product after wheat and corn. During the past two decades, the total yield of citrus products worldwide has increased by 60%, with approximately a three fold increase in China compared with Brazil and the USA. In the global market, the citrus trade essentially focuses on citrus processing products, which occupy one-third of the annual production. The citrus processing rate is approximately 70% and 85% in the USA and Brazil, respectively.

Y. Shan (Ed): Canned Citrus Processing. http://dx.doi.org/10.1016/B978-0-12-804701-9.00001-0

Citrus originated in China and has been cultivated there for 4000 years. In 2011, China was considered the largest citrus cultivator in the world, with approximately 2.28 million hectares for citrus cultivation, which produced approximately 29.4 million tons of citrus. Almost 182 million people in China are involved in the plantation of citrus, and the plantation areas are located in Hunan, Guangdong, Jiangxi, Guangxi, Sichuan, Hubei, Fujian, Zhejiang, and Chongqing provinces (Figure 1.1).

The plantation area in the Hunan province was approximately 379,000 ha and produced a yield of approximately 3.89 million tons of citrus in 2011; the plantation area and yield of citrus accounted for 16.5% and 12.8% of the total area and yield, respectively. Citrus fruits are primarily consumed as a raw food in China. Thus, the citrus processing rate is only 10%, and the main processed product is canned citrus segments.

The Hunan province is also responsible for manufacturing half of the processed citrus products and for exporting 60% of Chinese citrus products. There are currently 29 processing companies in Hunan. Sixteen of these companies are the leaders in Hunan, including Hunan Chic and Hunan Huaihua Huiyuan, the largest canned product processing factory and juice processing factory in China, respectively, and four of these companies are leading enterprises in China.

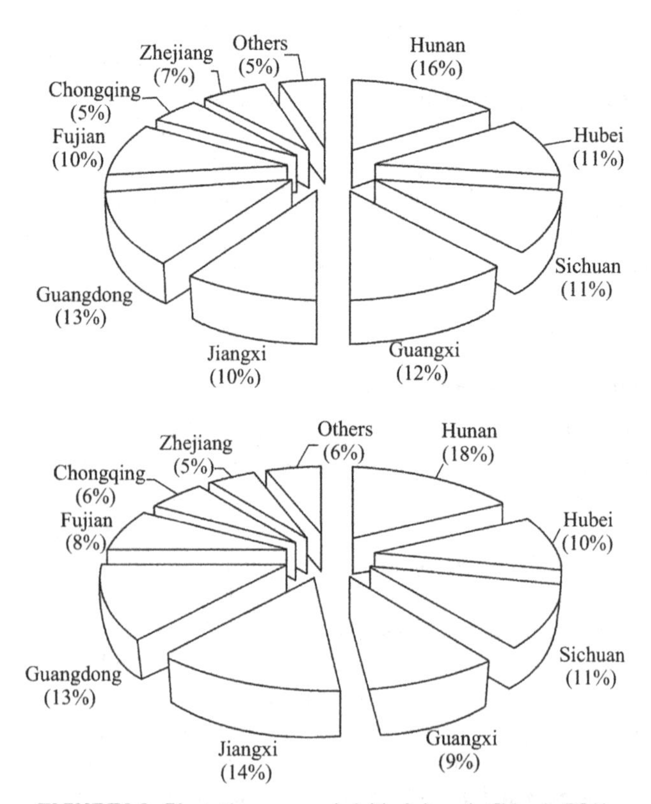

FIGURE 1.1 Plantation area and yield of citrus in China in 2011.

1.2 HISTORY OF DEVELOPMENT

Japan produced and exported the greatest amount of canned citrus products until China surpassed Japan in the mid-1980s. At present, China and Spain are the main producers of canned orange slices. Several advances in food processing technology improved and modernized canned citrus production. Automation of peeling, segment splitting, sac coating removal, and canning improved canned citrus processing. In addition, in the late 1980s, sterilization applied at low temperatures to canned orange slices greatly improved the quality and texture of the slices. Moreover, the application of commercially sterilized packages instead of traditional low-temperature storage significantly improved the quality of the canned slices.

Since the 1960s, canned citrus products were the major agricultural products exported from China, and this trend has not changed since then. There are three stages in the development of the canned citrus industry.

1.2.1 Starting Stage (1959–1969)

The first canned food factory was set up in Huangyan, Zhejiang Province, and had a production capacity of approximately 2000 tons per year. Later, small canned production factories were sequentially established in Hunan, Sichuan, Hubei, and other provinces. However, the market share of canned citrus products in China was less than 1% in the global market because the production of Satsuma mandarin was limited.

1.2.2 Development Stage (1970s to the Mid-1980s)

Plantations for Satsuma mandarin were quickly developed, and this drastically improved the productivity of the canned citrus industry. Annual exports improved to 12,000–20,000 tons after Satsuma mandarin was cultivated. However, outdated technology and instruments did not allow China to compete in the global market and high deficits plagued the canned citrus products during exportation. Market competition affected the advantages and disadvantages of the products; thus, only one-third of the factories survived and developed into backbone enterprises.

1.2.3 Prosperous Stage (1990s to the Present Day)

Japan, Spain, and other major countries have reduced their production of canned citrus products, while China has implemented novel technologies and equipment from other countries to improve the processing of canned citrus products. China's raw materials, price, and technological advances in food processing have increased the exportation of canned citrus products and caused Japan to become an importer rather than an exporter of these products. The Japanese and U.S. markets imported canned citrus products from Spain; however, China took over those markets. China has also placed large amounts of canned citrus products on the European market. Thus, China has become the production center of canned citrus products worldwide. At present, the available statistics reveals that over 0.8 million tons of canned citrus products are annually produced in China and that the Chinese produce more than 75% of the canned citrus products in the world. China exports more than 0.3 million tons of canned citrus products, and this accounts for over 70% of the canned citrus products in the world trade market. In 2010, approximately 0.34 million tons of canned citrus products were exported from China, and the export value was 280 million US dollars (Tables 1.1 and 1.2).

TABLE 1.1 Monetary value of exporting Chinese canned citrus products to major countries and regions

	Year 2010		Year 2009			
Country/ Region	**Export volume**	**Export value**	**Export volume**	**Export value**	**Amount increase (%)**	**Value increase (%)**
USA	169,147	151.45	167,947	145.08	0.71	4.39
Japan	52,137	46.05	48,585	42.00	7.31	9.64
Germany	32,860	24.29	28,970	21.61	13.43	12.38
Thailand	19,998	14.14	19,110	13.25	4.64	6.71
Netherlands	10,415	8.03	9559	7.50	8.96	7.08
Canada	9117	9.13	8905	8.59	2.39	6.26
United Arab Emirates	7609	3.55	6822	3.36	11.53	5.61
United Kingdom	5508	4.92	4326	3.57	27.33	38.07
Iran	4589	2.22	3380	1.49	35.75	48.91
Saudi Arabia	4217	1.58	5127	1.83	−17.75	−13.98
Korea	3415	2.50	3053	2.14	11.85	16.92
Tunisia	2441	1.43	1230	0.70	98.48	102.90
Czech	1954	1.55	1660	1.23	17.66	26.25
Yemen	1714	0.87	1350	0.85	26.94	2.86
Malaysia	1186	0.66	1362	0.69	−12.95	−4.34
Philippine	1092	0.93	1143	0.79	−4.46	17.94
Indonesia	706	0.51	143	0.08	393.25	506.06
Poland	702	0.52	641	0.44	9.46	19.35
Separate Customs Territory of TPKM	671	0.48	634	0.41	5.82	18.18
Australia	652	0.78	515	0.49	26.64	58.25
Other countries	6115	4.74	5895	4.15	3.73	14.17
Total	**336,244**	**280.34**	**320,359**	**260.26**		

Export volume is expressed in tons. Export value is expressed in million U.S. dollars.
Data source: China Canned Food Industry Association, 2011. The statistics of the value of exporting various canned products to major countries and regions on 2010. Food Canning(2), 63–68.

TABLE 1.2 Monetary value of exporting canned citrus products from major production areas in China

	Year 2010		Year 2009			
Areas	Export volume	Export value	Export volume	Export value	Amount increase (%)	Value increase (%)
Zhejiang	170,545	157.20	175,947	154.37	−3.07	1.83
Shandong	5353	4.27	4434	3.13	20.72	36.31
Hubei	81,890	49.42	72,144	43.13	13.51	14.58
Hunan	62,148	54.26	52,218	46.10	19.02	17.70
Guangxi	6160	4.40	4424	3.01	39.24	46.30
Anhui	7380	8.18	7283	7.17	1.33	13.97
Sichuan	1066	0.46	1458	0.64	−26.91	−28.21
Total	**334,542**	**278.19**	**317,908**	**257.55**	**5.23**	**8.01**

Export volume is expressed in tons. Export value is expressed in million U.S. dollars.
Data source: China Canned Food Industry Association, 2011. The statistics of export breeds and value from major production areas in China on 2010. Food Canning(2), 69–72.

1.3 CURRENT INDUSTRIAL SITUATION

Canned citrus products account for more than 80% of all processed citrus products. In China and in the global market, these products are traditional exports and the most competitive products, respectively. The Chinese canned citrus industry has recently entered into a stage of stable development and intends to improve product quality. At present, more than 0.8 million tons of canned citrus products are produced each year, and over 0.34 million tons of these products were exported in 2012, which produced an export value of 440 million U.S. dollars (Figure 1.2). These data reveal that the canned citrus products from China are responsible for 75% of the total international trade so that China has become the largest exporter of canned citrus products worldwide.

In recent years, China has greatly improved canned citrus processing techniques through (1) chelators and automatic flowing tanks, which remove the sac clothing and maintain the taste and 96% of the shape of the citrus segments; (2) application of continuous rotary sterilization under low temperatures, which avoids the generation of cooked flavor by dimethyl sulfide and keeps the citrus segments crispy; (3) using stabilizing reagents and enzyme inhibitors to resolve the white precipitation caused by hesperidin; and (4) reducing packaging costs and eliminating the fragile issue of glass bottles when ethylene-vinyl alcohol (EVOH), a degradable material with an antioxidative layer, was used in soft cans.

Recently, the Chinese canned citrus industry made several novel innovations: (1) the programmable logic controller water-saving system was based on an optimization theory and was developed to reduce water consumption. Water consumption for the canned citrus products decreased from 60–80 to 40 tons. (2) The enzymatic peeling or sac coating removal technology removed heavy

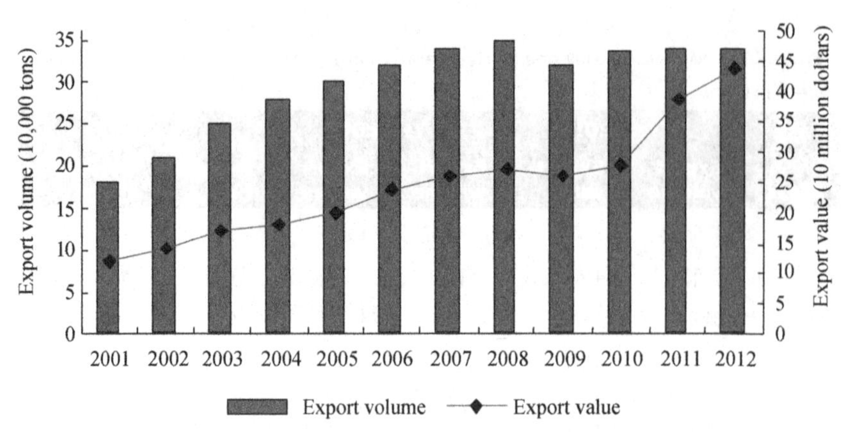

FIGURE 1.2 Export volume and export value of citrus cans (2001–2012).

metal residues (technical barrier) and eliminated acidic and basic wastes from the water generated during the canned citrus processing procedures (green barrier). Thus, this technological advancement improved the quality and safety problems associated with canned citrus products. (3) The quick freezing preservation technology preserves the original taste of citrus fruits, extends the raw material supply from 3 months to 1 year, and decreases packaging costs in the citrus processing industry.

Chapter | Two

Canned Citrus Processing

CHAPTER OUTLINE

2.1 Background 8
2.2 Categories of Canned Foods 10
2.3 Preservation Principles of Canned Foods 11
- 2.3.1 Effect of Exhausting on Preservation of Canned Products 11
- 2.3.2 Effect of Sealing on the Preservation of Canned Products 12
- 2.3.3 Effect of Sterilization on the Preservation of Canned Foods 13

2.4 Procedures and Techniques Used for Manufacturing Canned Citrus Products 18
- 2.4.1 Flowchart of the Process 18
- 2.4.2 Key Points of the Procedure 18
- 2.4.3 Presealing and Exhausting 27
- 2.4.4 Sealing 30
- 2.4.5 Sterilization 32
- 2.4.6 Cooling 35

2.5 Common Quality Issues and Controlling of Canned Foods 36
- 2.5.1 Swollen Cans 36
- 2.5.2 Corrosion of the Can Walls 37
- 2.5.3 Food Discoloration and Food Spoilage 37
- 2.5.4 Turbidity and Precipitation in Cans 38

2.6 New Sterilization Techniques of Canned Foods 38
- 2.6.1 Microwave Technique for Food Sterilization and Preservation 39
- 2.6.2 Ohmic Sterilization 39
- 2.6.3 Pulsed Light Sterilization Technology 40
- 2.6.4 Ozone Sterilization Technique 41
- 2.6.5 High-Pressure Sterilization 42
- 2.6.6 Pulsed Electric Field 43
- 2.6.7 Electrostatic Sterilization 44
- 2.6.8 Magnetic Sterilization 44
- 2.6.9 Induction Electronic Sterilization 45
- 2.6.10 Other Sterilization Techniques 45

Y. Shan (Ed): Canned Citrus Processing. http://dx.doi.org/10.1016/B978-0-12-804701-9.00002-2

2.1 BACKGROUND

In the past century, canned foods have become popular in the international market because they are easy to carry, convenient to eat, and have a long shelf life. Since the 1980s, more than 45 million tons of canned food were produced worldwide and more than half of those cans contained fruits and vegetables. China exports more than 10 million tons of canned fruits and vegetables, which are worth approximately 12 billion US dollars. In 2006, Chinese customs and other related organizations reported that 4.07 million tons of canned products were produced, 2.29 million tons were exported, and 2.35 billion US dollars was the export value. The reports also revealed that 1.36 million tons of the canned products were canned fruits and that approximately half of the canned fruit (0.60 million tons) were exported. Thus, compared with 2005, the export value for canned fruits in 2006 was 438 million US dollars, which resulted in a 7.78% and 18.84% increase in the export amount and export value, respectively.

The Chinese canned food industry started in 1906 when the first factory, the Shanghai Taifeng Food Company, was founded. Later, several small canned food factories were established in other provinces. Modernization of the canned food industry emerged during the Korean War to fulfill the food needs of the Chinese Volunteer Army. At that time, numerous methods were developed by Mr Zhang Xueyuan and his colleagues from the National Industry Department to eliminate the problems of food shortage that the army faced. Meanwhile, factories were established to produce canned meat products. The supply of canned foods greatly benefited the army and helped them win the war; thus, a large number of canned food factories materialized after the war. From 1949 to 2006, 59.49 million tons of canned products were produced; 25.43 million tons were exported; and 22.9 billion US dollars was the export value.

There are three main development stages in the Chinese canning industry after the establishment of PRC. The first stage started in the 1970s when canned foods were considered luxurious products for the average person. At the end of this stage, there were approximately 400 types of canned products. The National Department of Light Industry organized experts and technicians from the canned food factories, universities, institutes, and designing companies to compile standard techniques to produce canned products for the canning industry.

The second development stage started in the late 1980s and continued through the 1990s. During the transition process from a planned economy to a market economy, there were some disputes regarding the brands and trademarks of the canned products. The exported canned products with counterfeit trademarks affected the reputation of the Chinese canned products worldwide. Thus, it was difficult for the Chinese canning industry to advance in the international exporting market. Changes in the domestic and foreign markets affected the canning food industry, and several entrepreneurs were unsure of its future. However, since the mid-1990s, the Chinese canned food industry changed its economic growth pattern and diversified its structure. Thus, excellent quality and low prices of products from the Chinese canning industry, and the several canned foods from China caused it to gradually become the leader of the canning industry in the world market.

In the twenty-first century, Chinese canned products have moved into the rapid development stage, which is also known as the third development stage. At this stage, the production of canned products had substantially improved; in 2000, the total production of canned products increased by 11.3% (1.78 million tons) compared with the previous year. The total amount of Chinese canned products in the world market was 1.4 million tons, with 1.1 kg of canned product consumed by each person. In 2005, the total production of Chinese canned products was more than 3.6 million tons; 2.05 million tons were exported and 1.6 billion US dollars was the export value. The total sales in the Chinese canning industry were 26.9 billion RMB, which improved by 20.51% compared with the previous year. The profit of these canned product companies was 780 million RMB, which revealed a 41% increase in their profits. The countries that import the Chinese canned products are spread across five continents and include more than 140 countries or areas. The Japanese, USA, European Union, Russian, Association of Southeast Asian Nations, and Middle East markets receive the exported Chinese canned products.

The worldwide canning industry has developed into a large modernized industrial department. The USA, Italy, Spain, France, Japan, and the UK are major producers of canned foods. Approximately 50 million tons of canned foods are produced worldwide, and the average annual consumption for an individual is 10 kg (90, 23, and 1.6 kg for the USA, Japan, and China, respectively). There are over 2500 types of canned foods.

Based on the available statistics, the annual production of canned citrus products in China was more than 0.8 million tons, which is 75% of the total production in the world. More than 0.3 million tons of canned citrus were exported from China, and this was more than 70% of the global trade amount. In 2011, 0.337 million tons of canned citrus were exported from China, and the export value of these canned citrus products was 380 million US dollars. The only country that competes with China in exporting canned citrus products is Spain; however, Spain uses automatic instruments for the entire canning procedure and this alters the shape of the citrus slices. The automated process used in Spain for its canned products does not achieve the same high quality as that achieved by the manual peeling process used in China. Thus, Spain's exports of canned citrus products decreased as China's exports of canned citrus products gradually increased. Canned citrus products from the Hunan Chic Food Company were sold to the USA, Japan, Germany, Korea, and other countries. After the Hunan Chic Food Company installed two automatic packing systems that used EVOH antioxidative materials for canned citrus products, the product quality was found to as same good as that all of the other developed countries. The USA Food and Drug Administration (FDA) approved the exporting of canned products from the Hunan Chic Food Company to several countries and regions. Every year, there is a gradual increase in the amount of canned products exported by this company. In 2010, the

Hunan Chic Food Company collaborated with the Hunan Agricultural Product Processing Institute that specializes in agricultural product processing and developed the first automated production line in the world that removes the sac coating using enzymatic methods. This novel technique eliminates the pollution caused by traditional acid-base methods that remove the sac coating of the citrus fruits. In addition, removing the sac coating with enzymes improves the canned product quality and safety.

With the development of science and technology, new characteristics that improve the canning industry have emerged. The entire canning procedure has become automated; new techniques allow continuous production of canned products; and improved packing materials promote the safe consumption of the canned products. In addition, material storage is optimized, the pattern of canned products are transferred from general to specialized production, and the separation between empty can manufacturing and canned product preparation greatly improves production efficiency. Applying biological enzymes and fast freezing technology to the citrus slices as well as other modern techniques allows low energy consumption and less waste discharge in the canned product industry, as well as environment protection.

2.2 CATEGORIES OF CANNED FOODS

There are several types of canned foods and numerous categorization methods. Thus, in 1989, China published the categorization standard of canned foods (GB 10784—1989). In this standard, the canned foods were divided into six categories on the basis of raw materials. Each category was further divided into several subcategories according to the processing methods or taste. The categories below pertain to the canned citrus products.

(1) Sugar solution cans: The peel and seeds are removed from the citrus fruits, and then the fruits are placed in the cans. Different concentrations of sugar solution are added to the cans to produce the canned citrus products. These products are known as canned citrus with sugar solution. The most popular canned product in the global market is the canned citrus slice without the sac coating in sugar solution.

(2) Syrup cans: The sugar is boiled until 60–65% of the soluble solids are dissolved. Citrus materials and high-concentration syrups are added to the cans to create the canned citrus products with syrup. Canned kumquat products with syrup are an example of syrup cans.

(3) Canned sauce: Canned citrus products are classified into one of the following categories on the basis of product recipes and requirements.

a. Jelly: Pretreated fruits are boiled with or without water, and the fruits are squeezed, juiced, separated, filtered, and clarified. Then, sugar, citric or malic acid, pectin, and other additives are added to the semifinished products. The mixed solution is condensed into final products with 65–70% of soluble solids. The final product is packed into cans and is called jelly, which can be divided into three subtypes.

i. Pure jelly and fruit jelly: This type of jelly is made by boiling one or more mixed juices and concentrating it after adding sugar and citric acid in a constant ratio.

ii. Pectin jelly: This jelly is mixed with water, tartaric acid, sugar, pectin, and other additives in a constant ratio.

iii. Pectin fruit jelly: This jelly is a mixture of fruit jelly and pectin jelly.

b. Marmalade: This product uses citrus fruits as raw materials. Several exocarps are cut into filaments and immersed in sugar solution until the exocarps are clear in color. The pretreated exocarps are added to the citrus fruits (raw materials) as an additive and are homogeneously distributed. Marmalade has bitter and sweet tastes.

(4) Canned juice: Citrus fruits are homogenized, squeezed, and filtered to obtain the juice that is sealed in metal containers. The canned juice can be divided into three subtypes according to the product quality requirements.

a. Original juice: Fresh fruits are squeezed to produce this 100% fresh fruit juice. This juice is not fermented or condensed, and its appearance is transparent or cloudy.

b. Fresh juice: Sugar and citric acid are added to the original or concentrated juice to obtain a 30% minimum concentration.

c. Concentrated juice: The original juice is boiled to obtain one to six times concentrated juice.

2.3 PRESERVATION PRINCIPLES OF CANNED FOODS

The enzymatic reactions that occur in food spoil the food. These reactions are caused by the growth and proliferation of microorganisms. However, foods that are sealed and preserved in cans have a prolonged shelf life. Exhausting, sealing, and sterilizing of canned citrus products kill the microorganisms that cause putridity, toxicity, and diseases and inactivate the enzymes in the raw materials. Sealing the cans prevents microorganisms from contaminating the cans.

2.3.1 Effect of Exhausting on Preservation of Canned Products

Exhausting is a technique that releases air from food. Specifically, air generated by the canning process, air in the histiocytes of raw materials, and air that is present in the cans' headspace is released, and this creates a vacuum space in the headspace after the can is sealed. The most common exhausting methods are thermal exhaust, vacuum sealing, and steam injection sealing. The vacuum degree is between 20 and 45 mm Hg.

Excessive oxygen in the canned products causes adverse changes such as spoilage, reduced quality, and corrosion of the inner wall of the cans during storage. Exhausting is beneficial because it inhibits the growth of aerobic bacteria and fungi, alleviates inner wall corrosion caused by the canned foods, and eliminates the changes in the color, smell, and taste of the canned foods. Exhausting also reduces the destruction of vitamins and other nutrients and prevents can deformation due to air expansion during sterilization.

(1) Effect of exhausting on microorganisms: Canned foods are commercially sterilized; however, few microorganisms remain alive in the canned foods after sterilization. These microorganisms are usually aerobic bacilli that require sufficient oxygen for their growth. During the exhausting procedure, air in the canned products is eliminated and the oxygen concentration is greatly decreased. Thus, this procedure effectively prevents the growth of aerobic bacilli, and thus inhibits the spoilage of canned foods and extends the shelf life of these foods.

(2) Effect of exhausting on food quality and nutrition: When food is exposed to air, its surface is easily oxidized and its taste, color, shape, and nutrition are affected. Citrus jam, jelly, and fruits in sugar solutions change in terms of color, smell, and taste.

(3) Effect of exhausting on the corrosion of the inner wall of the cans: The inner walls of the cans are easily corroded during storage. This occurs because of an electronic chemical reaction, which is decided by an anode and a cathode. The corrosion speed is affected by several factors. If oxygen is present in the can, it will depolarize the cathode and corrosion will be accelerated. This is especially true for canned fruit products with high acid concentrations; oxygen may accelerate the corrosion of the metal foil and even cause perforation. The corrosion of the inner wall of the cans decelerates after the oxygen concentration is reduced by exhausting.

(4) Effect of exhausting on the appearance of the cans: Exhausting maintains the pressure inside the cans at a lower level than the outside pressure, which results in the concave bottom of the cans. When food is spoiled, it becomes rancid and produces gas. This increases the pressure and reduces the vacuum inside the can. In severe situations, cans become swollen with protruded bottoms. The quality of these canned products is determined by a dull solid thumping sound or a hollow resonance that occurs when someone knocks on the bottom of the can.

2.3.2 Effect of Sealing on the Preservation of Canned Products

Canned food has a long shelf life because the sterilized food is separated from the outside environment by the sealed containers (i.e., metal can, glass can, and aluminum foil). Thus, food is not contaminated with outside air or microorganisms. The sealing efficiency directly affects the shelf life of the canned products. Thus, high-quality sealing causes the canned food to have a long shelf life, while low-quality sealing causes the food to have a short shelf life. Most canned foods are sealed using a sealing machine, and a strictly controlled sealing procedure is vital for food storage.

2.3.3 Effect of Sterilization on the Preservation of Canned Foods

Because sterilization determines the shelf life of the cans, it is an essential method in manufacturing canned products. Sterilization kills microorganisms in canned foods by heating or by using other methods. However, sterilizing canned food does not make the food bacteria-free but prevents pathogenic bacteria or viruses from remaining in the food. It is allowed that certain microorganisms or spores remain in the food because the food is not spoiled in certain period while the microorganisms are in a special environment of the can.

Heating, flame sterilization, and radiation sterilization are techniques used to sterilize canned products. Heating is the most widely used one.

1. Effect of sterilization on microorganisms in canned foods

Canned food contains several microorganisms; however, only pathogenic bacteria or bacteria that cause food spoilage are killed. Microorganisms have an optimal temperature range that is imperative for their growth and replication. However, once the temperature exceeds the optimal range, bacterial growth is inhibited or the bacteria are killed. Canned foods have a long shelf life because heating kills the microorganisms by denaturing their cellular proteins and destroying their metabolic enzyme activities. However, the sensitivity of microorganisms to heat depends on the type and number of microorganisms, environment, and heating conditions. Different sterilization methods are used for different microorganisms.

(1) Spoilage and spoilage microorganisms in canned foods: Microorganisms that can cause food spoilage and change the quality of the canned foods are called spoilage microorganisms. Based on the differences in food types and characteristics as well as the procedures and storage conditions of canned foods, the spoilage microorganisms could be fungi, yeasts, molds, or a combination of these microorganisms. The type of spoilage microorganisms occurring in different canned foods with various acidities also depends on the optimal acidity required by these microorganisms for growth.

Based on the differences in bacterial tolerance to different pH and heating conditions, canned foods are commonly divided into four types: low-acidic, middle-acidic, acidic, and high-acidic foods. Canned citrus belongs to the acidic or high-acidic food groups.

In the canning industry, low-acidic and acidic foods have a pH boundary of 4.6. Low-acidic foods are sterilized because the bacterium *Clostridium botulinum* that produces the botulinum toxin (Botox), which is harmful to humans, grows in environments with a pH greater than 4.6. Thus, this bacterium must be eliminated from canned foods with low acidity. Furthermore, acidophilus bacteria, such as *Bacillus stearothermophilus*, are resistant to heat and commonly survive in low-acidic foods. Thus, these bacteria require additional techniques to eliminate them from the low-acidic foods.

Clostridium pasteurianum is a common bacteria found in acidic foods. The spoilage microorganisms in highly acidic (pH=3.7) foods include bacteria, yeasts, and molds that are highly tolerant to acids and minimally tolerant to heat. However, in highly acidic canned foods, enzymes have higher tolerance to heat than the spoilage microorganisms. Thus, inactivating the enzymes is critical for sterilizing highly acidic foods.

(2) Heat tolerance of microorganisms: The type and number of microorganisms, as well as the environment, affect the heat tolerance of microorganisms. Sterilization of canned products involves application of heat to food to kill microorganisms. The proteins in the cells of these microorganisms are denatured and their metabolic enzyme activities are inactivated. Thus, protein stability primarily determines the heat tolerance of microorganisms; however, acidity, alkalinity, and salt and water content also affect the heat tolerance of microorganisms.

a. Factors that affect the heat tolerance of microorganisms.

i. Different bacteria and different strains of the same bacterial species tolerate heat differently. Thermophilic bacteria have the highest tolerance to heat, forming more heat-tolerant spores than vegetative cells. Many types of bacteria, especially *Bacillus*, are involved in contaminating food. Thus, to effectively kill bacilli, these bacteria should undergo the same sterilization process as that used for other bacteria; however, pasteurization should occur for a longer time.

ii. Medium and food components influence pasteurization. Acidity affects food more than any other food component. However, bacilli have the highest tolerance to heat in a neutral environment, and its heat tolerance dramatically decreases at a pH lower than 5. Thus, addition of acids to the food increases the foods' acidity during processing, reduces the sterilization temperature, shortens the sterilization time, and maintains the original quality and taste of the food.

Addition of acid to decrease the heat tolerance of microorganisms reveals that different acids change the food differently. Lactic acid has the strongest inhibitory effect on microorganisms, followed by malic and citric acid. Foods with a low pH (acidic or highly acidic) may be sterilized at lower temperatures (<100 °C), at low pressure, and for shorter sterilization times, while foods with a high pH (or low acidity) need to be sterilized at higher temperatures (>100 °C), at high pressure, and for longer sterilization times.

In addition to acids, the heat tolerance of microorganisms is affected by sugar, salt, proteins, and phytoncides. Longer sterilization times are required to kill bacilli at higher sugar concentrations, while shorter sterilization times are required at lower sugar concentrations. High sugar concentrations protect bacilli, and this protection may be due to the use of sugar by these bacilli to dehydrate the plasma of the cells, which influences denaturation of proteins and the heat tolerance of the bacterial spores. If *Escherichia coli* is sterilized at 70 °C, then the time required to eliminate these bacteria from 10% sugar solution will be extended to an additional 5 min compared with that needed to eliminate these bacteria from a sugar-free solution. A 30% sugar solution prolongs the sterilization time by 30 min. However, when the sugar concentration is very high, environmental osmotic pressure is also high, thus greatly inhibiting the growth of microorganisms.

The heat tolerance of bacilli is protected when the salt concentration is less than 4% but weakened when the concentration is over 8%. However, the protection effect changes depending on the type of spoilage bacteria. Permeation of low salt concentrations causes cells to partially absorb water, making it difficult to denature proteins, and enhances the heat tolerance of microorganisms. High osmotic pressure, resulting from high salt concentrations, causes large quantities of proteins to dehydrate, leading to the death of microorganisms. Salt solutions with Na^{+}, K^{+}, Ca^{2+}, and Mg^{2+} have antibacterial effects that are dependent on temperature. Salt can also decrease the water activity (A_w) of food by reducing the amount of water available for microbial metabolism and thus inhibiting microbial growth.

Heat tolerance of microorganisms can be directly or indirectly influenced by starches, proteins, and fats found in food. While starch does not directly influence bacilli, gelatin and albumin, which are found in proteins, as well as fats and oils, increase their heat tolerance.

iii. For microorganisms at certain concentrations, their lethal conditions are also determined by heat treatment temperature and time. Chemical reactions are always accelerated at high temperatures. Thus, increasing the temperature accelerates protein denaturation and reduces heat tolerance. Previous experiments have confirmed that lethal time of microorganisms during thermal sterilization is exponentially reduced with increased temperatures.

iv. With an increase in the bacterial number, the pasteurization time needed for spoilage bacteria and their spores also increases. Thus, it is important to minimize the number of bacteria present in food prior to sterilization.

b. Common parameters for heat tolerance in microorganisms

i. *F* value: The *F* value is the heating time (in minutes) required for killing microorganisms at a specific lethal temperature and a constant concentration. In general, the lethal time at 121 °C is also called sterilization efficiency, lethal value, or sterilization intensity. During the design of the sterilization program, the strongest spoilage bacteria and toxin-producing microorganisms are killed to determine their heat tolerance. For example, the lethal time for *C. botulinum* is 2.45 min at 121 °C in a phosphate buffer (pH 7.0).

F value includes safe (or standard) and actual sterilization values. The safe *F* value is based on ideal conditions with an instant temperature adjustment, and it is referred to as the standard for sterilization conditions. To obtain this value, microorganisms in the canned products are detected and the most common bacteria, which cause contamination, are selected. The types and number of common bacteria that cause contamination is then used to simulate the *F* value. However, a temperature increase or decrease process occurs during actual production. During this process, sterilization efficiency is accomplished under the lethal temperature. Thus, the experimental procedure can be based on the safe *F* value and heat conductivity. Measuring the inside temperature change for canned products determines the actual *F* value. Normally, the actual *F* value is slightly higher than the safe *F* value. If the actual *F* value is lower than the safe *F* value, then sterilization will be incomplete and the sterilization time or temperature will be increased. If the actual *F* value is considerably greater than the safe *F* value, it will result in excessive sterilization. In this situation, we would shorten the sterilization time or decrease sterilization temperature to ensure the quality of the canned products.

ii. *D* value: The *D* value is the time required to kill 90% of the original bacteria in certain environments and under lethal temperatures. It is the time for one cycle in the log curve under lethal temperatures. For a specific bacterium, the lethal time varies with the heat treatment temperature. Higher temperatures result in shorter sterilization times. Under experimental conditions, a constant bacilli concentration is examined using the lethal temperature and time. The results are described using a semilogarithmic plot, which is also called a thermal lethal temperature–time curve, with the lethal temperature plotted on the x-axis and the lethal time plotted on the y-axis. The *D* value is highly correlated with heat tolerance such that a high *D* value indicates that the microorganism has a high heat tolerance. If the *D* value is high, then the sterilization time required to kill 90% of the bacilli will be prolonged.

iii. *Z* value: The *Z* value is the temperature required to shorten the lethal time by 10-fold. For example, if the *Z* value equals 10, the lethal time will be shortened to 1/10 of the original value and the sterilization temperature is increased by 10 °C. A high *Z* value indicates that the microorganism has a strong tolerance to heat. In the thermal lethal temperature–time curve, the reciprocal of the absolute value of the slope represents the *Z* value.

iv. TDT value: The TDT value is the lethal time required for all microorganisms under a specific temperature.

(3) Effect of heat conductivity of canned products on sterilization efficiency: During pasteurization, canned products with low temperatures constantly capture energy from a heating source (i.e., steam or boiling water). The temperature of each spot in the canned product increases as heat accumulates. In general, the slowest rate of heat absorption occurs in the center of the canned products. Heat is gradually transferred from the outside to the inside of the can. In the cooling procedure, heat is gradually transferred from the inside to the outside of the can from a cooling media (i.e., water or air). The heat-transferring patterns and the time it takes for various canned foods to absorb the heat are different and are affected by many factors. The heating degree of each spot inside the canned products differs, suggesting that various canned food or even different positions of the same canned product are affected differently by sterilization. Thus, the heat conductivity of canned food is a critical factor in determining a suitable sterilization procedure.

a. Major heat-transferring procedures for canned foods

During pasteurization, food properties and sterilization methods cause the heat-transferring pattern to vary. Heat-transferring patterns commonly include conduction, convection, and radiation. Heat conduction and convection are the major heat-transferring procedures.

i. During the heating or cooling process, heating temperatures cause molecules to generate different vibrating energies. When molecules collide, conduction transfers heat energy from high-energy molecules to low-energy molecules. Temperature differences between the medium and the food during the heating or cooling process causes a temperature gradient, which is generated between the interior wall and the geometric center of the can. In both the heating and cooling processes, heat is transferred from high-temperature spots to low-temperature spots (i.e., from the geometric center to the can). Thus, each spot in the can is heated differently, and the cold spot, which is the slowest spot to heat and cool, determines the geometric center. Because food is a poor conductor of heat, it takes a long time for the temperature to reach the cold spot by heat conduction. Thus, a longer pasteurization time is required for solid or viscous canned foods to reach the cold spot.

ii. Convection uses liquid or gas flow to transfer heat: The sterilization time is shortened because heat convection occurs quickly, and this rapidly alters the temperature in the cold spot. Convection is the heat-transferring pattern for canned foods supplemented with salt solution, sugar solution, or any other low viscosity solution.

b. Factors of canned foods that affect heat transfer

i. Physical properties associated with canned food: Shape, size, concentration, viscosity, and density are the physical properties of food that are influenced by the heat transfer pattern and rate. Solid or highly viscous foods have low heat conduction because the foods are stationary inside the can during the heating process. Sugar, starch, jelly, and other ingredients in liquid foods also affect the heat transfer rate and patterns during pasteurization.

ii. Initial temperature: The initial temperature of the food represents the temperature at the beginning of sterilization and the temperature before the retort is heated. The FDA states that when thc heating process is about to begin, the temperature of the retort is based on the temperature after the first sealing. A high initial temperature decreases the difference between the initial and sterilization temperatures and shortens the time required for the sterilization temperature to reach the center of the food. Thus, the high initial temperature is critical for canned foods requiring sterilization using a heat conduction procedure.

iii. Containers: The heat resistance of a container affects the heat transfer rate, which varies with different types of containers. Glass containers are highly heat resistant, while metal containers are not. Thus, heat transfer is slower for glass containers than for metal containers, and the sterilization time is longer for glass containers than for metal containers. Furthermore, the sterilization time is longer for large containers than for small containers because small containers have larger surface area per unit volume to transfer heat.

iv. Types of sterilization instruments: Retorts (rotary and stationary) are sterilization instruments. During sterilization of canned foods, rotary retorts are continuously swirling, which makes heat transfer faster and more consistent in these retorts than in stationary retorts. Rotary retorts promote heat conduction of fluid food or food with poor fluidity. For example, canned citrus products in sugar solution have a faster sterilization speed if rotary retorts are used rather than stationary retorts.

v. Other factors: Many other factors influence heat transfer. Can filling, headspace, vacuum degree, liquid-to-solid ratio in the can, fruit maturity, processing methods of raw materials, food features during heating (e.g., formation of precipitation at the bottom), temperature distribution inside the can before heating, stacking during canning, heat distribution of sterilization retorts, and length of heating time affect heat transference and will be discussed below.

2. Effect of sterilization on enzyme activity in canned foods

Organisms contain enzymes, which are specific proteins and biological catalysts. Fruits and vegetables contain various enzymes that accelerate the decomposition of their organic compounds. Controlling enzyme activity is crucial in order to avoid adverse effects on raw materials and products. Thus, controlling enzyme activity prevents the food from turning into low-quality food and preserves the nutrition of the canned food. Enzyme activity is closely related to temperature; catalytic velocity is doubled if the temperature increases by 10 °C. The optimal temperature for optimal enzyme activity ranges from 30 to 40 °C. If the temperature reaches up to 80 °C, most enzymes will be irreversibly damaged by the heat. Enzymes are deactivated by altering molecular surfaces, breaking chemical bonds or rings, and destroying protein structures.

During manufacturing, certain enzymes cause deterioration of acidic or highly acidic foods. Moreover, heat used during sterilization can reactivate enzymes such as peroxidases because at high temperatures, enzymes remain active after microorganisms have perished. Thus, canned fruits or vegetables with high acidity need additional treatment because the internal temperature is not high enough to deactivate all of the enzymes. For example, pasteurization protects canned grapefruit juice against bacterial spoilage; however, pectinesterase remains active. Enzymes are easily deactivated under humid conditions but not under dry conditions. To completely deactivate enzymes and avoid or decrease the deterioration caused by the enzymes, comprehensive methods need to be considered.

2.4 PROCEDURES AND TECHNIQUES USED FOR MANUFACTURING CANNED CITRUS PRODUCTS

2.4.1 Flowchart of the Process

Flowchart of innovative technology of citrus cans is shown in Figure 2.1.

2.4.2 Key Points of the Procedure

1. Precleaning raw citrus fruits

Washing separates soil, stones, microbes, pesticide residues, and leaves from citrus fruits. In certain cases, the citrus fruits are cleaned using a multi-stage wash. Sometimes cleaners containing hydrochloride acid, citric acid, and detergents such as bleach and potassium permanganate are involved in the washing process. The four steps in the washing process are flushing, soaking, rinsing (water spraying), and high-pressure water spraying.

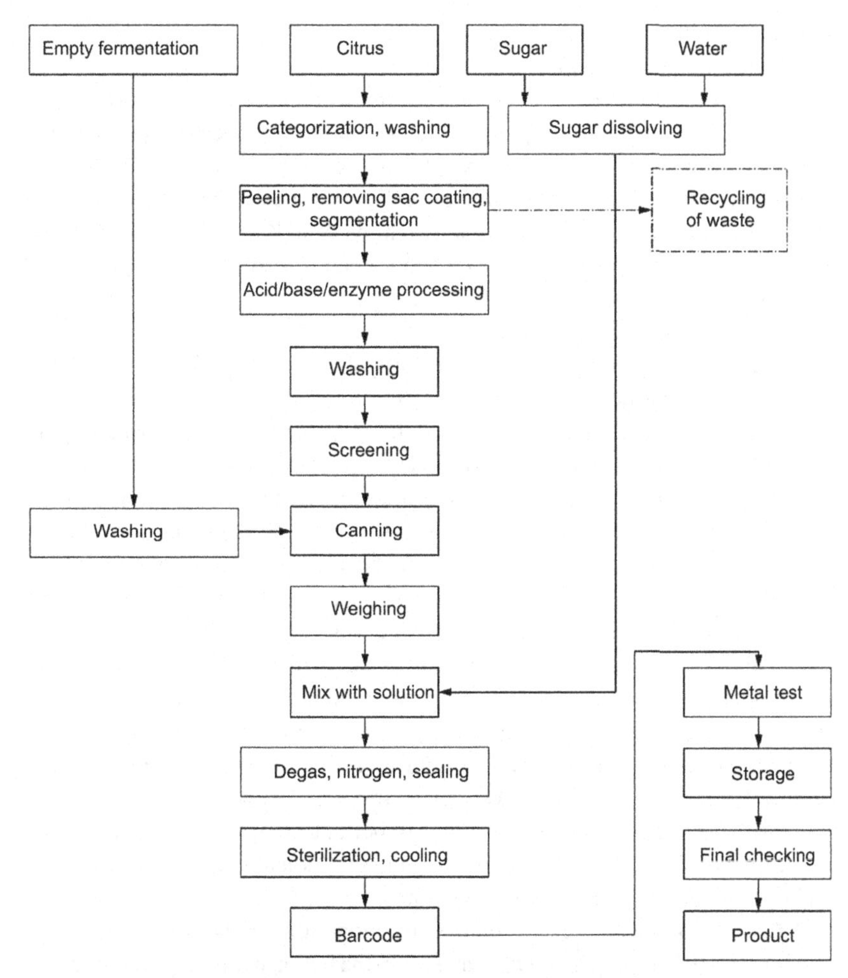

FIGURE 2.1 Flowchart of innovative technology of citrus cans.

(1) Flushing is performed in an open or closed water flume, which is positioned at a certain slope. Citrus fruits are washed with flowing water in this step.

(2) Soaking occurs after the citrus fruits are hoisted to a water sink.

(3) The rinsing process begins with sorting the fruits. Branches, leaves, decayed and immature fruits, as well as fruits infected with bugs or those that have any other defects are removed so that the quality of the product is not affected. After sorting, the citrus fruits are cleaned using a washing machine equipped with multibrush rollers. Rinsing occurs when water is sprayed above the fruits and the fruits are washed with rollers that have brushes.

(4) The final step in the washing process is high-pressure spraying. Water used in all the washing steps is filtered and disinfected. Thus, water used in the washing process can be recycled.

2. Peel and pith removing

1) Traditional techniques

Citrus fruits are transported to the blanching chamber by a platform elevator. During the transportation process, the blanching chamber sprays the citrus fruits with steam at 5–10 bar. Usually, citrus fruits are blanched for 30–60 s, and the blanching depth is 1–2 mm under the citrus fruit peels. After blanching, the peels burnt by the hot steam are manually removed, and the pith is removed. The integrity of citrus segments needs to remain intact so that the overall quality of the segments is maintained.

2) Enzymatic peeling technique

The citrus peel is composed of an exocarp (also known as the oil cell layer or colored cortex), mesocarp (also known as the sponge layer or white cortex), sac coating, juice sac, seeds, and the juice in the sac. Citrus fruits are usually covered by a thick cuticle, which can cause inconvenience during consumption and processing. The citrus peel is usually removed manually. However, this method is labor-intensive and time-consuming and, in most cases, the white cortex is not removed entirely. Furthermore, citrus segments are often damaged during processing, which leads to product loss. Thus, peeling of citrus fruits has received a lot of attention. There have been reports that citrus peels are removed by peeling devices and freeze–thaw cycles. At present, most studies focus on the enzymatic peeling technique, which needs to be further examined in both research and industrial fields. Chinese peeling studies are limited to the blanching process. Recently, the Hunan Agricultural Product Processing Institute revealed a simple method that removes peels of fresh citrus fruits and preserves the nutritional components of the fruits. This research is important to the Chinese citrus industry because it allows the industry to remain internationally competitive.

(1) Preparation of the enzyme solution

A 0.4% (V/V) pectinase and cellulase reaction mixture (the ratio of pectinase to cellulase is 2:1) is prepared with water set to the national drinking water quality standards.

(2) Pretreatment of fresh citrus fruits

After rigorous washing, the citrus fruits are soaked in a cold enzyme reaction mixture (3–10 °C) with a ratio of 1 part citrus to 1.5–2 parts enzyme solution for 15–20 min in a vacuum degree of 20–30 mm Hg.

(3) Enzymatic digestion

Citrus fruits are digested in a 40–50 °C water bath for 30–40 min.

(4) Treatment after enzymatic digestion

After digestion, the citrus fruits are washed, packed, and preserved in the refrigerator.

(5) Results

Citrus peels are successfully removed by enzymes after the pretreatment procedures. The results indicate that the exocarp and mesocarp are removed under the following conditions: the citrus to enzyme solution ratio adjusted to 1 part citrus to 1.5–2 parts enzyme solution, the pectinase to cellulase ratio of 2:1, enzyme concentration of 0.4% (V/V), enzymatic reaction time of 40 min, and temperature of 45 °C. The optimal pH is adjusted and controlled. The data also showed that enzymatically peeled citrus fruits are identical to fresh fruits in color and taste and that their nutritional components are preserved. For example, enzymatic peeling preserved 97% of the vitamin C in citrus fruits, which satisfied the demands of the consumers.

3. Removal of the sac coating

In the citrus food industry, a vital processing step is the removal of the sac coating of citrus fruits while the orange segment syrup or canned citrus is prepared. In China, the acid-base two-step method, the modified two-step method, the phosphate-sodium hydroxide (NaOH) one-step method, and the weak alkaline method with ethylenediaminetetraacetic acid (EDTA) as the additive are the most commonly used techniques for removing the slackcloth.

1) Conventional method

The classical method for removing the sac coating is the acid-base two-step method. Citrus segments are treated for 40–50 min with 0.1–0.2% hydrochloric acid (HCl) followed by 0.05–0.3% NaOH for 5–10 min. The citrus segments are rinsed and washed with water for 30–60 min.

The theory behind the conventional acid-base two-step method can be explained as follows. Acids destroy the sac coating while an enormous amount of naringin is dissolved. Alkaline solutions are also applied to the citrus fruits to eliminate the sac coating and dissolve hesperidin, which prevents turbidity and white precipitation in canned citrus products.

(1) Disadvantages of the conventional acid-base two-step method

Limonin and naringin are components of citrus fruits. The acid-base two-step method uses hesperidin to offset the effects of naringin; however, this method fails to neutralize limonin. While naringin exists in certain types of citrus fruits (e.g., grapefruit), limonin analogs are more common in citrus plants. Experimental results have confirmed that limonin is not found in citrus fruits; however, its precursor, limonin A-ring lactone (LARL), is present in these fruits. In the canning industry, LARL is released when fresh citrus fruits are damaged. During the acidic phase of this method, LARL is converted to limonin and a bitter flavor is generated in the canned citrus products. The bitterness can be reduced by extending the washing and rinsing time. However, the time required to remove the bitterness decreases the nutritional value of the citrus fruits. Extensive washing also causes the fruits to absorb water, and this complicates the manufacturing process.

(2) The theory of removing the sac coating using EDTA as an auxiliary chemical

Polysaccharides such as pectin and cellulose are found in the sac coating of citrus fruits. Removing the sac coating essentially destroys the pectin–cellulose complex by converting protopectin to soluble pectin. Destruction of the pectin–cellulose complex allows the sac coating to be removed by mechanical force.

Protopectin is composed of pectin molecules that are cross-linked through salt bridge formations between the free carboxyl groups and bivalent cations such as calcium (Ca^{2+}) and magnesium (Mg^{2+}). Removing Ca^{2+} and Mg^{2+}

converts protopectin to soluble pectin while NaOH or HCl disrupts pectin macromolecules by hydrolyzing the α-1,4-glycosidic bonds. After the pectin–cellulose complex breaks, the ruptured sac coating is removed by mechanical force such as a high-pressure air blast.

Many types of chemicals including EDTA, polyphosphate, and citrate can chelate Ca^{2+} and disrupt the sac coating by breaking the Ca^{2+} bridge. EDTA is the most powerful chelating reagent. It does not cause water pollution and chelates Ca^{2+} and Mg^{2+} in water, which enhances the ability of NaOH to disrupt pectin macromolecules. Thus, utilizing EDTA is more advantageous than using polyphosphate.

(3) NaOH and EDTA remove the sac coating of citrus fruits

After the citrus fruits are graded and blanched, the segments are separated and treated with 0.04% EDTA-Na_2 and 0.15% NaOH for 15 min. The EDTA-Na_2 and NaOH-treated segments are rinsed to remove alkali. Then, the segments are soaked in 0.1% citric acid for 20–30 min. The sac coating removal procedure is concluded by rinsing off the remaining acids. High-quality canned citrus fruits are obtained if the fruits are sequentially treated with alkali and then acids to remove the sac coating. Alkali treatment represses LARL conversion to limonin and neutralizes the carboxyl group (—COOH) of LARL by removing it. In addition, alkali treatment removes hesperidin and acid soaking removes naringin. The naringin concentration determines the soaking time and temperature. All of these substances that alter the flavors of the canned citrus products can be removed with alkali, EDTA, and acid treatment. The soaking process removes naringin. Soaking time depends on the naringin concentration in different citrus species, while the increase in soaking temperature significantly improves the bitterness.

2) Enzymatic removal of the sac coating in the citrus fruits

The canned citrus industry uses microbial or biological enzymatic techniques to remove the sac coating. Mechanical devices and freeze-thaw methods have also been used to remove the sac coating. The most popular technique, from all of the aforementioned techniques, is the use of bioactive enzymes, which is a safe, efficient, reproducible, and environment friendly method. Researchers from the Hunan Agricultural Product Processing Institute investigated the optimal conditions for removing the sac clothing using enzymatic methods. The sac coating of fresh and mature citrus fruits, especially fruits of mandarin orange, were removed using compound enzyme preparation under optimal conditions. The results from this study provide a theoretical foundation and a significant value for the future industrialization of citrus fruits.

(1) Principles for enzymatic removal of the sac coating

Pectin and cellulose are found in the sac coating of the citrus segments. Utilizing enzymes to degrade pectin and cellulose results in a disruption of the sac coating, and this allows the sac coating and juice sac to be separated without damaging the orange pulp. Citrus segments are enveloped by a thin translucent membrane structure called the sac coating. The sac coating has 10–20 layers of parenchyma cells. Parenchyma cells in the inner layer are larger and have an

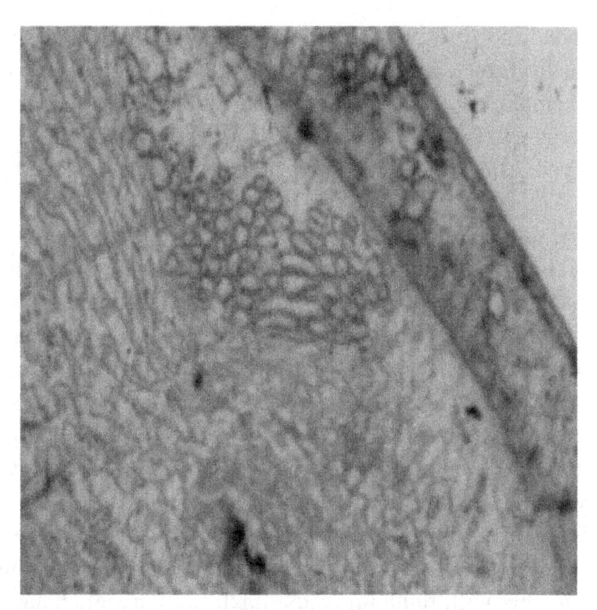

FIGURE 2.2 Electron microscope image of the natural segment membrane.

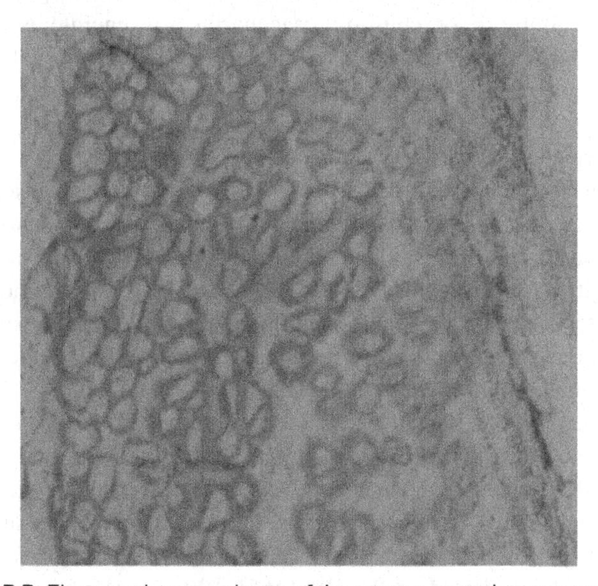

FIGURE 2.3 Electron microscope image of the enzyme-treated segment membrane.

irregular alignment, while cells in the cuticle layer are smaller and are regularly and tightly aligned. The exterior layer is coated with wax. Cells are connected to pectin and cellulose; the structure is observed using an electron microscope in Figure 2.2.

Figure 2.3 illustrates the effects of compound enzymes on the sac coating, which was evaluated using an electron microscope. The figure shows that gaps between the sac coating cells are enlarged after enzymatic treatment. The cuticle layer is irregular and disordered. Enzymes are highly sensitive and specific; thus, they usually react with their specific substrate. If a compound enzymes preparation is employed to disrupt the pectin–cellulose complex, then intercellular connections will be lost, structures will become loose, and the sac coating will be destroyed. After the sac coating is destroyed, juice sacs can be separated from the sac coating.

(2) Enzyme selection

Biological enzymes are highly specific, and certain compound enzymes recognize specific substrates. To optimize the conditions for sac coating removal, compound enzymes were screened. Several compound enzymes have distinct functions that depend on the chemical compositions of the sac coating and on the types and ratios of these enzymes. For instance, cellulase contains at least three types of enzymes: endocellulase (EC 3.2.1.4, Cx enzyme or CMC enzyme), exocellulase (EC 3.2.1.91, CBH, C1 enzyme or micro-crystalline cellulase), and β-glucosidase (EC 3.2.1.21). Endoglucanase randomly hydrolyzes β-1,4-glycosidic bonds at amorphous sites, cleaves long cellulose chains, and produces considerable amounts of small cellulose fragments with reducing ends. Each time, exoglucanase cleaves a single cellobiose from the ends of a linear cellulose chain. β-Glucosidase hydrolyzes cellobiose or cello-oligosaccharide to generate glucose, and it exhibits high enzymatic activity on trisaccharide and disaccharide substrates. As the degree of polymerization of glucose increases, the hydrolysis rate decreases.

(3) Enzymatic hydrolysis conditions in a pilot scale study

The Hunan Agricultural Product Processing Institute conducted a pilot study that was based on single factor tests and optimization of compound enzymatic hydrolysis conditions. The study examined the enzymatic removal of the sac coating of citrus fruits using devices for peeling and sac coating removal. At pH 4.5, the factors such as enzyme concentration, reaction time, temperature, and electrical frequency were selected to optimize the conditions for the enzymatic removal of the sac coating of the citrus fruits.

The results indicate that an enzyme concentration of 0.35% at 40 Hz (electrical frequency) causes hydrolysis at pH 4.5 for 50 min and at 45 °C. These data revealed the optimal enzymatic conditions for removing the sac coating of the citrus fruits. Citrus segments subjected to the enzymatic removal of the sac coating have good appearance (i.e., tight texture and excellent gloss), improved vitamin C retention (based on the GB 11671—2003 test), and reduced metal residues (GB 5175—2000) compared with those subjected to the conventional sac coating removal method. Thus, the enzymatic method of sac coating removal is better than the conventional acid–alkali method.

4. Canning

1) Washing empty cans

Microbes and other contaminants, which are introduced during manufacturing and transportation, are removed when the cans are washed and disinfected. Metal cans and new glass cans are usually cleaned with hot water, while recycled cans are cleaned with 2% alkaline water (or other detergents) and rinsed with disinfected fresh water. Cleaned and recycled cans have to be used immediately to avoid recontamination.

2) Preparation of infusion liquids

Infusion liquids or syrups are commonly used in canned vegetables and fruits, especially as syrups for canned citrus. However, these liquids are not used in juice or fruit jam. In the can, infusion liquids fill the space among the citrus fruit, improve flavor, eliminate air, and enhance the initial temperature and heat transfer.

(1) Preparation of the syrup

The syrup concentration depends on the citrus species, fruit maturity, amount of pulp loaded into the can, and quality of the product. In Chinese sugar citrus cans, the glucose concentration must be 14–18% when the can is opened. The following formula was used to calculate the syrup concentration:

$$w_3 = (m_3 w_2 - m_1 w_1)/m_2$$

where

m_1 represents pulp-loading mass in each can (g);

m_2 represents syrup mass in each can (g);

m_3 represents the net weight of the can (g);

w_1 represents the mass fraction (%) of the soluble pulp solid substance;

w_2 represents the required syrup concentration when the can is opened (%);

w_3 represents the concentration of the prepared syrup (%).

In the canning industry, syrup concentrations are measured using a refractometer or saccharometer. Because temperature affects the density of a liquid, syrup is usually standardized to 20 °C. The syrup is composed of sucrose with 99% purity. Granulated sugar and water are weighed, mixed, boiled in a sugar dissolving pot, and then filtered. The syrup is usually made as a concentrated (65%) stock solution, and it is then diluted with water or a diluted solution when it is being packaged. The syrup is boiled for 5–15 min to diminish microbes and to remove sulfur dioxide (SO_2), which alters the color of the citrus fruits. Immediate filtration after boiling is often employed.

(2) Salt solution was prepared using refined salts, which contain more than 98% sodium chloride (NaCl). Refined salt is weighed, dissolved in water, and boiled.

3) Canning requirements

Citrus fruits are canned shortly after the syrup is added in order to prevent microbial contamination of the semifinished products. A sorting step is essential to ensure the quality of each citrus can. The standard sorting protocol removes rotten or diseased citrus slices to maintain slices with uniform size, color, and shape.

In addition to quality, quantity is important in the canning industry. Each can varies in weight because of the different citrus species and types of cans. However, both net weight and solid content have to meet the standard requirements. Net weight is defined as the total weight of an item (i.e., weight of the solid citrus fruit and syrup). It is calculated by subtracting the weight of the empty can from the total weight of the final product. The solid content or the net content is defined as the weight of the solid citrus slices in the can with the required syrup concentration (45–65%). A standard for the net content depends on the size of the cans and the type of canned citrus fruit. The standard deviation for the net content is ±3%, but the average is greater than or equal to the standard. The net content of manual or machinery canning was examined by random sampling.

Semifinished products of different raw materials tend to vary slightly; thus, it is important to maintain uniform color, maturity, segment size, and number in each can during the canning process.

During the canning process, the headspace, which is the space between the food and the lid of the can, should remain empty. The headspace reveals a direct effect on loading weight, sealing performance, deformed or pseudo-swollen (nonseptic) cans, metal corrosion, color change, and spoilage of canned foods. During sterilization, a small headspace causes pressure to increase inside the cans, leading to microbial growth because of the adverse effects on the sealing process. Limited headspace also causes permanent deformation or lid protrusion from the metal cans. Metal corrosion produces hydrogen, which accumulates inside the can and causes the can to swell; this swelling impairs its sale. However, if the headspace is too large, then there will be no sufficient content in the can and excessive oxygen will fill the space in the can. Excessive oxygen corrodes and deflates the cans and oxidizes the citrus products in the can, leading to a change in the color or quality of the canned citrus products. The headspace requirement is typically 4–8 mm between the top surface of the food and the rim of the can. Other impurities should be removed during the canning process to ensure the quality of the product.

4) Canning method

Manual and mechanical canning are the two canning methods that are generally used. Selecting a canning method depends on the type of food and canning requirements. In China, manual canning is used for citrus fruits that do not have the sac coating surrounding the fruit. This canning method involves loading, weighing, and adding the syrups. The entire procedure is conducted on an operating platform supplemented with a conveyor belt that transports materials, empty cans, and filled cans. This method is simple and is extensively applied to various canned foods. Manual canning keeps the citrus segments intact, while mechanical canning does not. However, manual canning involves greater content deviations among the cans, low canning efficiency, and poor sanitary conditions. Manual canning lacks continuity in the canning industry.

Mechanical canning is commonly used for semisolid or liquid foods (jam and juice, respectively). This method is accurate, clean, and can easily adjust the amount of content in each of the cans. Thus, this method is highly efficient and has continuity in the canning industry; however, it is limited in its use in canning small-scale trail products.

2.4.3 Presealing and Exhausting

1. Presealing

Prior to sealing the cans, certain canned foods are presealed. Presealing occurs when a cap is loosely placed on a can by a sealing machine. The loose cap can be rotated on the can, but it cannot be removed easily. Presealing ensures that air is released from the can without the cap being removed during thermal exhausting and vacuum sealing. This method prevents food from spilling out of the can because of thermal expansion, and it prevents water of condensation from entering the food and contaminating it during the exhausting process. Furthermore, presealing prevents food from being damaged because of direct heating on its surface. It also stops cold air from entering the can, keeping the temperature and vacuum degree stable. Presealing also ensures that there is a good fit between the can and the cap. Thus, presealing improves the quality of the seal, especially for square-shaped or other odd-shaped cans.

2. Exhausting

1) Function of exhausting

Exhausting occurs when air is expelled from the materials and headspace. This procedure occurs after canning and is vital for the canning process. Exhausting allows canned fruits to attain a certain temperature (using a vacuum degree), which decreases after the cans are sealed and disinfected. Thus, exhausting helps maintain the quality of the canned citrus fruit. Details of the exhausting process are summarized below.

(1) Exhausting prevents high-temperature sterilization from deforming the cans or damaging the canned citrus products. If exhausting is omitted, food and air will expand while water inside the cans will evaporate, leading to a rapid increase in the pressure inside the cans. If the pressure is greater than the limit of the can, the double seam will become loose and the cans will become pseudo-swollen or protruded. Moreover, the can could explode and the cap could pop off of the glass can. Omission of exhausting causes the aforementioned effects, leading to defective canned products. However, if exhausting is utilized, almost all of the air will be released from the can, thus reducing the pressure inside the can and resolving the aforementioned issues.

(2) Exhausting also prevents the growth of aerobic bacteria and molds. Canned food is subjected to commercial sterilization, which means that live microorganisms are limited but still exist in sterilized canned foods. Aerobic bacteria belonging to *Bacillus*, which require adequate oxygen to support their growth, are the most prominent microorganisms. Thus, exhausting inhibits the proliferation and growth of bacilli by removing oxygen from the can and extends the storage life.

(3) Exhausting does not allow oxygen to alter the color and taste of canned citrus. When food and oxygen are in a can, the top layer of the food inside the can is oxidized, leading to a change in the color, smell, and taste of the food. Exhausting reduces the oxygen concentration in the food, water, and juice inside the can so that vacuum is generated and oxidation of the food is prevented.

(4) Exhausting attenuates the destruction of vitamins and other nutritional components. During manufacturing, nutritional components of canned foods can be damaged depending on the type of food, heating temperature, heating time, and oxygen concentration. Because exhausting reduces the oxygen in the food or headspace, vitamins and other easily oxidized nutritional components are somewhat protected.

(5) Exhausting prevents or alleviates the corrosion of inner walls of the cans during the storage period.

(6) Exhausting is helpful for quality inspections because the quality of canned food can be identified by looking at the indentures on the bottom of the can. The vacuum degree is usually inspected by the sound produced after hitting the bottom of the cans with a small stick. The sound changes if the food inside the can is spoiled due to a decrease in the vacuum degree. These cans can be easily sorted.

2) The vacuum degree of canned foods

The vacuum degree is defined as the difference in the pressure outside and inside the cans. This difference should be between 26.7 and 40 kPa. The more the air is in the can, the lower is the vacuum degree. And higher vacuum degree results from less air existence in the can. The exhausting process determines the amount of air that remains in the cans. High vacuum degrees cause the bottom of cans to become flat or indented. Canned foods with high quality tend to have this prominent exterior feature, and this is often used to determine the quality of canned products.

3) Factors related to the vacuum degree

The vacuum degree determines the exhausting efficiency after the sterilized canned foods have cooled. Higher vacuum degrees indicate that exhausting was very effective. Exhausting temperature, acidity and air content in the food, canning condition, headspace size, and exhausting method affect the vacuum degree of the canned foods.

(1) Exhausting time and temperature: During the high-temperature exhausting process, an increase in exhausting temperature and time causes a high vacuum degree. High temperature results in the quick expansion of air in the can, and adequate heat allows the food to expel the air outside the can. Extending the heating time ensures sufficient removal of air. However, overheating and prolonged exhausting cause the pulp to soften and rot and the syrup to overflow. In addition, a high vacuum degree, after sealing, leads to deflated cans.

(2) Headspace size: Headspace is a crucial factor for the vacuum degree. A larger headspace generates a high vacuum degree; however, a considerably large headspace has adverse effects. At each temperature, there is a threshold for the headspace. If the headspace threshold is below the vacuum degree, then the vacuum degree will increase with increasing headspace. However, if the headspace threshold is higher than the vacuum degree, then the vacuum degree will decrease with increasing headspace. The maximum vacuum degree is obtained at threshold. The headspace threshold is not constant and can be increased by increasing the temperature. Studies on the headspace threshold have gained tremendous attention from many researchers in the packaging industry.

(3) Food acidity: The vacuum degree is correlated with the acidity of canned foods. Highly acidic foods erode the metal walls of the cans, leading to hydrogen production. The hydrogen produced increases the pressure inside the cans, further decreasing the vacuum degree. Because citrus fruits are very acidic, paint-coated cans are highly recommended to prevent erosion and maintain a high vacuum degree.

(4) Variation in atmospheric temperature: The vacuum degree is defined as the difference in the pressure between the inside and outside the cans. At high atmospheric temperatures, air expansion in the cans cause the vacuum degree to decrease. Thus, vacuum degrees tend to deteriorate when processed cans are transported from a place with low temperatures to a place with high temperatures.

(5) Variation in atmospheric pressure: The vacuum degree is related to atmospheric pressure; it decreases at low atmospheric pressure. Atmospheric pressure decreases as the altitude increases; thus, transportation of cans from a low-altitude place to a high-altitude place decreases the vacuum degree.

4) Exhausting methods

Thermal exhaust, vacuum sealing exhaust, and steam sealing exhaust are the three exhausting methods commonly utilized in the canning industry. The oldest exhausting method is thermal exhaust and is still utilized by many industries. Vacuum sealing exhaust was invented after thermal exhaust and is the most commonly used method. Steam sealing exhaust was recently developed and is not extensively used because it is relatively new to the domestic canning industry compared with the two other methods.

(1) Thermal exhaust removes air from the cans using the basic principles of thermal expansion for food and air. Heating causes the food and air to expand and water to partially vaporize inside the can. The increased water vapor pressure removes air from the can. Heat sterilization, cooling, and exhausting of the cans were immediately followed by sealing of the cans, which generated a certain vacuum degree due to food shrinkage and water vapor condensation.

The thermal exhaust method can be divided into hot can exhaust and heating exhaust in an exhaust cabinet.

Hot can exhaust heats the can to a certain temperature and then seals the can immediately. This method is suitable for liquid and semiliquid foods or foods that have components and morphology that cannot be stirred or damaged, respectively, during heating, such as citrus jam. When using this exhaust method, it is vital to maintain the food inside the can at a high temperature. If the temperature does not meet the required standard, then the canned foods will not reach the expected vacuum degree. After the canned foods are sealed, immediate sterilization is required; otherwise, thermophilic microorganisms will grow and reproduce at this temperature. The reproduction of these microorganisms will cause the canned foods to deteriorate if sterilization does not occur immediately. Reheating is required if the average temperature of the filled can is lower than the required standard temperature.

Heating exhaust is the other exhausting method. Canned foods that may or may not be capped are delivered to an exhaust cabinet, which is heated to a certain temperature to ensure that the temperature at the center of the cans meet the industrial requirement (usually 80 °C). After air is completely expelled, the cans are immediately sealed and sterilized, which results in a specific vacuum degree inside the cans after the cans are cooled.

The heating exhaust temperature and duration depends on the type of food, size of the empty cans, type of cans (metal or glass), and form of the food (liquid, semiliquid, or solid). Exhaust temperatures usually range from 90 to 100 °C, while the heat exhaust duration is 5–20 min. Heating exhaust effectively removes air and partially functions as a sterilizer. However, heating exhaust is a time-consuming technique that adversely affects the color, smell, and taste because of high temperatures and has a low efficiency.

A chain-belt exhausting cabinet and a geared-disc exhausting cabinet are the apparatus used for heating exhaust. Steam injection pipes are present on both sides of the bottom of the chain-belt exhausting cabinet. The amount of steam injected is controlled by valves to maintain a certain temperature in the exhausting cabinet. On the driving force of a chain belt, the cans are transported from one end to other end of the exhausting cabinet. Then, the exhausted cans are discharged from the steam cabinet. The only difference between the chain-belt exhausting cabinet and the gear-disc exhausting cabinet is the driving style of transportation. The gear-disc exhausting cabinet uses geared discs to drive the transportation of the cans.

(2) The vacuum sealing exhaust method uses a vacuum can sealing machine. Cans will be exhausted and sealed while the vacuum can sealing machine is vacuuming the cans. This method provides the cans with a high vacuum degree of 35–40 kPa or greater.

The vacuum sealing exhaust method can quickly increase the vacuum degree in a very short time. This method preserves vitamins and other nutritional components, exhausts various canned products, and is widely utilized in many canning industries because the machine uses a small amount of space. However, this method can only be used for exhausting the headspace because of its limited exhausting time. Furthermore, the vacuum sealing exhaust method has difficulties removing air from food, especially food with high air content. Food with high air content needs to be vacuumed before it is canned. When using this method for exhaust, the vacuum degree in the sealing machine and the temperature of the food in the cans should be taken into consideration in order to avoid food overflow during the sealing process.

In addition to the aforementioned exhaust methods, a steam exhaust method injects steam into the headspace of the can to repel air and then sealing immediately follows.

2.4.4 Sealing

Sealing is one of the most important factors that affect the long-term storage of canned food by preventing environmental bacteria from contaminating the food. To ensure that cans and caps are tightly sealed, sealing machines are used to seal the cans. It is critical that the sealing procedure is strictly controlled. The sealing methods and requirements will vary depending on the types of cans used.

1. Sealing of metal cans

(1) Composition of the sealing machine: The semiautomatic, automatic, and vacuum sealing machines are the three types of can sealing machines utilized in the canning industry. A can sealing machine is composed of pressure heads, trays, and rolls. Pressure heads are used to keep the cans stationary, preventing the cans from moving and sliding while they are being sealed and thus ensuring the quality of seaming. Cans are kept on trays so that the pressure heads can be embedded into the caps. Rolls are round-shaped pinions composed of abrasion-resistant, high-quality steel. There are two types of rolls: primary rolls with deep pressure-relieving grooves (i.e., first rolls) and secondary rolls with shallow pressure-relieving grooves (i.e., second rolls). The rolls possess distinct structures and functions. Primary rolls fold the end curl of the lid around the flange of the can to form the structure, while secondary rolls iron out the seam into the final shape.

(2) Types of seam sealing: In brief, it is made of a magnesium alloy plate. A rubber band or a polyvinyl chloride plastic ring is used to ensure the tight fit between the lids and cans. Steam spraying or vacuuming is usually applied for sealing the can and for generating vacuum inside the cans. Then, a clamp sealing machine tightens the connection between the lid and the body of the can.

2. Sealing of soft canned products

During manufacturing, the inner edge of composite plastic film layers can be connected or closely fused together to achieve the sealing requirements. However, composite plastic films are not completely sealed no matter how good the plastic film is. Other methods are based on the properties of the composite plastic films and the food packaging status. These sealing methods include the high-frequency, hot-press, and pulse sealing methods.

(1) High-frequency sealing method: This method produces a seal that has enormous strength and good tightness. However, water or oil on connection surfaces can impair its sealing effects. Therefore, this method is appropriate for sealing soft cans but not for sealing food after it is packaged.

(2) Hot-press sealing method: A metal heat sealing rod is covered with polytetrafluoroethylene on its surface to form a protective layer. This is an electrical rod and reaches a certain temperature to melt the inner layer of the bags and fuses both films together under pressure. The melting degree depends on the material properties of the composite plastic film bags, temperature, time, and pressure of the material to be melted. To improve the sealing strength, a cold press was applied after the heat sealing procedure. This sealing method can be used when the inner surfaces are covered with a small amount of water or oil. Thus, this method can be used to seal polyethylene, waterproof cellophane, and polyethylene composite films.

(3) Pulse sealing method: This method is similar to the hot-press sealing method, but the difference between the two methods is the heating source. When a high current flows through resistance wires, a large amount of heat is produced. This fuses and seals the inner plastic walls together. At a low voltage, the high-density current instantaneously passes through the fine resistance wires to cause an instantaneous increase in the temperature of the heating plate, which fuses the inner layers of the plastic bags with a small amount of water or oil. This method has the characteristics of a convenient operation because it is

widely applicable and has high melting strength and sealing strength, which is better than other sealing methods. It is the most commonly used method.

Simple heat sealing machines, vacuum or vacuum gas-filling heat sealing machines, pulse vacuum packaging machines, and gas exchange packaging machines are used to seal soft cans.

A height of 3.8 cm between the surface of the loaded food and the opening of the soft can needs to be maintained to ensure that the soft cans are sealed. Furthermore, after the soft can is sealed, a cooling period is required for the soft can before the can is moved from the vacuum chamber to an atmospheric environment.

2.4.5 Sterilization

Sterilizing canned foods kills and removes microorganisms such as pathogens, toxigens, and putrefactive bacteria and deactivates enzymes in the food. This method provides canned food with long-term storage (more than 2 years) and prevents spoilage. However, sterilization in the canning industry is different from microbiological sterilization. In the canning industry, sterilization does not completely eliminate all of the bacteria, but only pathogenic and spoilage bacteria are eradicated because these bacteria cause human diseases or food spoilage. Pasteurization, radiation sterilization, microwave sterilization, and high-pressure sterilization are the sterilization methods used in this industry. At present, the most commonly used sterilization method is pasteurization. Experimental data indicate that the efficiency of pasteurization is improved when the temperature increases. For example, the pasteurization efficacy at 120 °C is enhanced 100-fold compared with that at 100 °C. However, an infinite increase in the pasteurization temperature is unrealistic because high temperatures will severely damage the food by destroying its nutritional components and thus the quality of the food.

Pasteurization eliminates harmful microorganisms by heating the canned food. To select the appropriate sterilization method, a thorough understanding of the factors contributing to sterilization is required.

(1) Technical conditions of sterilization: Temperature, time, and counterpressure are the factors that affect sterilization of canned foods. In the canning industry, a sterilization equation is commonly used to explain the relationships among all factors. The sterilization equation is as follows:

$$(T_1 - T_2 - T_3)/t \text{ or } (T_1 - T_2)/t,\ P$$

where

T_1 represents the heating duration (min);

T_2 represents the sterilization time (the time required to keep the sterilization temperature constant) (min);

T_3 represents the time needed to cool down (min);

t represents the required sterilization temperature or the maximum temperature that the sterilization pot could reach (°C);

P represents the counterpressure inside the sterilization retort during the counterpressure cooling procedure (Pa).

There are three stages in sterilization; the heating-up stage, constant-temperature stage, and cooling-down stage. The heating-up stage is the time it takes the temperature to rise from the initial temperature to the sterilization temperature (constant temperature) in the sterilization retort. This stage plays important roles in expelling air from the retort and maintaining a constant pressure. The constant-temperature stage is the period of sterilization where the retort reaches the sterilization temperature while the cooling-down stage is the time that temperature decreases to the opening temperature after heating ceases. This stage also enables the pressure to be released from the retort.

Optimal sterilization conditions help eliminate all the harmful bacteria and deactivate the enzymes. Sterilization temperature and time are also critical in determining food quality and in storing the canned food.

(2) Sterilization methods: Aseptic canning and can sterilization are sterilization methods that are based on whether sterilization is conducted before or after sealing. In aseptic canning, food and cans are sterilized separately while canning and sealing occur in an aseptic environment. This method slightly affects the food quality. Can sterilization sterilizes the food and cans after canning, and this method is commonly used in canning industry.

Different sterilization methods are used depending on the raw materials and the types of cans. Specifically, there are three sterilization methods that are based on raw materials: atmospheric sterilization (temperature does not exceed 100 °C), high-temperature and high-pressure sterilization (temperature range is 100–125 °C), and ultrahigh-temperature sterilization (temperature is over 125 °C). Techniques and equipment are used to divide the sterilization methods into metal can sterilization, glass can sterilization, and soft can sterilization. The sterilization methods determine the equipment to be used.

a. Discontinuous high-pressure sterilization: This method can use steam or water bath sterilization. Stationary high-pressure pasteurization retorts are the main equipment used in this method. Sterilization occurs in a sealed horizontal or vertical container with high pressure and continuous stirring in a batch mode.

Appropriate equipment is necessary for sterilization. In high-pressure steam sterilization, the steam is evenly distributed inside the retort and heats the retort until it reaches the sterilization temperature in a certain time period. The air exchange rate enables the retort to reach the sterilization temperature after the air inside the retort is expelled.

b. Discontinuous atmospheric sterilization: Canned citrus fruits are usually sterilized by boiling the fruit at atmospheric pressure. Iron cages carrying canned citrus fruit are merged in the sterilization retort and boiled according to the sterilization period. The temperature difference between canned citrus and water has to be measured to avoid the expansion of the glass cans. However, high temperature and extended boiling could result in soft, rotten, and shattered citrus segments; thus, this method can reduce the color, flavor, and other food quality characteristics.

c. Continuous sterilization: This type of sterilization can use atmospheric, low-temperature, or hydrolock machineries.

i. Continuous atmospheric sterilization machinery: Canned foods are continuously transported to the machine by a conveyor belt that is sterilized by boiling water. The sterilization time is controlled by the speed of the conveyor belt.

ii. Continuous low-temperature sterilization: Canned foods are sterilized (for 12 min) with a continuous sterilization machine that is supplemented with dual water baths (85 °C, pH of 3.7) to ensure the sterilization efficiency. A number of thermophilic bacteria continue to grow even after the bacteria endure sterilization at 100 °C, but these bacteria are vulnerable to an acidic environment (low pH) that can be created by increasing the H^+ concentration. A low pH shortens the sterilization time and reduces the sterilization temperature. Thus, pH is a critical factor in low-temperature sterilization, and pH must be less than 3.7. This method preserves the color, flavor, and taste of the canned citrus products better than the conventional atmospheric sterilization. Low-temperature, rotary, and short-time sterilizations are better than high-temperature, stationary, and long-time sterilizations, respectively.

Experimental data show that low-temperature sterilization reduces the shrinkage level of the citrus segments, increases the amount of canning, and shortens the sterilization time compared with continuous atmospheric sterilization. The continuous low-temperature sterilization is popular because its high productivity leads more revenue.

iii. Hydrolock continuous sterilization: This method joins rotary sterilization with cooling, and it can sterilize iron cans, glass cans, and plastic bags. In brief, this method uses a conveyer belt to transport canned food into a high-pressure steam sterilization chamber using hydrolock valves. Cans are sterilized by steam at a constant pressure and undergo a horizontal reciprocating movement. The sterilization period is controlled by the conveyer belt.

Hydrostatic pressure sterilization uses three stages to sterilize the cans. A conveyer belt transports the cans to a steam heating chamber after the cans pass through a preheated hydrostatic column. Then the cans are sterilized in a steam heating chamber for a designated period of time, and finally, the cans are discharged from the steam chamber after the cans pass through the cool water. Temperature and pressure in the sterilization chamber can be adjusted by changing the height of the preheating and cooling hydrostatic water. The pressure can reach 0.147 MPa (equivalent to 126.7 °C) when the height of the hydrostatic entry leg is 15 m. Hydrostatic pressure sterilization is steam-saving, time-saving, and performs uniform sterilization. However, the machinery associated with this method is very space consuming and expensive. Thus, hydrostatic pressure sterilization is appropriate for those factories that manufacture products with similar hydrostatic treatments.

iv. "Shan Guang 18" sterilization: This method is an aseptic sterilization method that uses "Shan Guang 18" equipment. "Shan Guang 18" is composed of a cylinder-shaped compression cabin that is used for canning and sealing, a compression valve, a pressure-releasing valve and air locks. Canned foods subjected to high temperature, and short sterilization times are transported to a compression cabin for canning and sealing. Food cans remain in the cabin for an additional 5 min, and the 5 min will be used for cooking and sterilization prior to the cooling down process. This method has two advantages, a variety of common canning and sealing equipment can be employed inside the compression cabin, and it is not a necessity to autoclave the empty can or to create an aseptic environment for canning and sealing.

v. Ultrahigh-pressure sterilization: Canned foods are subjected to sterilization treatment (at 1000 atmospheric pressure), which inhibits bacterial growth and facilitates its long-term storage. Ultrahigh pressure is usually generated by a hydraulic device and not by a pneumatic device. The current manufacturing scale is limited to equipment with a high-pressure-generating pump in a discontinuous manner. To sterilize a large amount of food, containers with continuous batch processing devices are required. Canned foods repeatedly go through the following process: food input, pressure treatment, pressure maintenance, pressure release, and the discharge of food. The packaging materials and containers should be highly pliable, which allows the containers to regain its original form.

2.4.6 Cooling

After heat sterilization, the canned food should be cooled so that the color, taste, and quality of the canned food are maintained. Food in the can remains heated; thus, rapid cooling steps such as atmospheric cooling and counterpressure cooling need to be performed.

(1) Counterpressure cooling: This method is commonly used for cans that are sterilized with high pressure and high temperature, especially cans that are prone to deformation or damage after sterilization. Cans remain in the sterilization retort for a certain period of time to allow the cans to completely cool down. Cold water and compressed air are usually poured into the retort to maintain the pressure equilibrium until the pressure approaches atmospheric pressure.

(2) Atmospheric cooling: This method is commonly used for canned foods sterilized by the atmospheric sterilization procedure. The cans will be cooled down in the retort, soaked in a cold water bath, placed under running water, or sprayed with water. Spraying water is the most effective cooling method; it allows water to evaporate and the evaporation to absorb the heat, which completely and rapidly cools down the cans.

Cooling time depends on the type of food, the sizes of the cans, the sterilization temperature, and the temperature of the cooling water. The temperature required for cooling is 38–40 °C. The residual heat on the surface of the cans facilitates water evaporation from the cans and prevents the cans from rusting. Sanitation of the cooling water and the breaking glass cans are two important issues in the cooling process that require special consideration. The water used for cooling has to meet the drinking water standards. If chlorination is a necessity, then the concentration of free chloride in the cooling water should be maintained between 3 and 5 mg/kg. Breaking glass cans can be minimized by gradually cooling the glass and closely monitoring the cooling temperature.

2.5 COMMON QUALITY ISSUES AND CONTROLLING OF CANNED FOODS

2.5.1 Swollen Cans

The bottom of qualified cans should be flat or indented. However, if the internal pressure is higher than the atmospheric pressure, then the bottom ends of the cans swell. This phenomenon known as swollen can is caused by overloading the cans with hydrogen and bacteria.

1. Physically swollen cans

(1) Causes: Overloading the cans and limited headspace causes the cans to swell. Food inside the can expands during heat sterilization, and a swollen can emerges after the can is cooled down. Furthermore, quick pressure and fast cooling after heat pressure sterilization, inadequate exhaust, or a very high storage temperature can easily result in swollen cans. The physically swollen cans contain food that is in a good condition and is still edible.

(2) Prevention: Physical swelling can be prevented by controlling the canning volume, generating a high vacuum degree, and gradually cooling the cans. The canning volume can be controlled by maintaining 3–8 mm for the headspace and utilizing adequate exhaust. The temperature at the center of cans should be increased to generate a high vacuum degree after the pressure sterilization process. The pressure inside the cans should be gradually reduced to the atmospheric pressure.

2. Hydrogen-induced swollen cans

(1) Causes: A hydrogen-induced swollen can is chemically induced and is caused by hydrogen production in the can. Hydrogen is released after the internal walls of the iron cans are corroded by acidic food. Then, hydrogen increases the pressure inside the cans and causes the cans to swell. Canned foods with hydrogen-induced swelling are still edible but do not meet the canned food standards. Thus, hydrogen-induced swollen cans are not recommended as edible foods.

(2) Prevention: In order to prevent hydrogen-induced swollen cans, damage to the internal walls of the empty cans should be avoided, thus reducing the direct contact between food and the exposed iron. The materials for the empty cans should be changed from iron to steel, and a coating of intact paint or highly acid-resistant paint should be used. In addition, the components and concentrations of acid in the food should also be monitored and adjusted.

3. Bacteria-induced swollen cans

(1) Causes: This type of swelling of the can is mostly caused by inadequate sterilization or sealing. The reinvasion of bacteria and the degradation of the food in the can occur when the cap lacks sufficient tightness to prevent gas production and increases the pressure in the can, thus resulting in a bacteria-induced swollen can.

(2) Prevention: Thorough washing and sterilization protects canned foods or semifinished products from contamination. Sufficient heating without damaging the nutritional components is also critical in preventing bacterial growth. The addition of organic acids, such as citric acid, to the preheated water or syrup could increase the pH level and repress bacterial growth. Controlling the sealing quality of the cans and the hygiene associated with the cooling water is vital in preventing bacteria-induced swollen cans.

2.5.2 Corrosion of the Can Walls

1. Influencing factors

(1) Oxygen: Oxygen is a strong oxidant to metal, especially in the acidic environment of the canned foods. The oxygen concentration in the cans is one of the critical factors that contribute to the corrosion of the walls.

(2) Acid: Citrus fruits contain high concentrations of acidic components, which is another important factor that contributes to the corrosion of the walls. The types of acids determine the amount of corrosion that will occur.

(3) Sulfur and sulfide: Sulfur or sulfide is an element that is present in many types of pesticides. It is also present in granulated sugar as an impurity. In canned foods, sulfur or sulfide as well as nitrates can cause the can walls to corrode.

2. Prevention

(1) To remove the pesticides, fruits are soaked in 0.1% HCl for 5–6 min and then rinsed.

(2) Foods that contain a lot of air can be vacuumed to reduce the oxygen concentration in the food.

(3) Extended heating improves the vacuum degree inside the cans.

(4) Syrup should be boiled to eliminate sulfur dioxide (SO_2) in the sugar. A coating of acid-resistant or anti-sulfur paint must be used for the containers that contain syrup.

(5) The humidity for storing canned food should be between 70% and 75%. This prevents acid or sulfur from corroding the external walls of the cans. Moreover, rust-preventive oil should be applied to the external walls.

2.5.3 Food Discoloration and Food Spoilage

Canned vegetables and fruits may change in color or taste during storage or transportation. Chemical reactions in the presence of enzymes, residual oxygen in the can, or high temperature can cause enzymatic and nonenzymatic browning of the vegetables and fruits.

Furthermore, residual flat-sour bacteria, such as thermophilic bacilli, in canned foods can produce sour flavors due to food spoilage. A bitter flavor in canned foods is caused by citrus pith or seeds.

Prevention:

(i) According to the canning requirements of various food, blanching with proper temperature and time deactivates enzymes and expels the air in the raw materials.

(ii) Syrup should be made fresh by boiling. If acid is required, it should not be added too early in the process because sugar could react with the amino acids to cause nonenzymatic browning.

(iii) The direct contact of fruits or fruit chunks with metals such as iron or copper during processing can be prevented by using stainless steel for the manufacturing of cans. Moreover, water used in the canning industry should not contain high concentrations of heavy metals.

(iv) To remove flat sour bacteria and prevent the spoilage of canned foods, the cans need to be thoroughly sterilized.

(v) Citrus pith and seeds should be completely removed; seedless citrus fruits are the ideal raw materials. The storage temperature of the warehouse needs to be controlled because storage at low temperatures or high temperatures could result in light browning or accelerate the browning, respectively.

2.5.4 Turbidity and Precipitation in Cans

During canning, high concentrations of Ca^{2+} and Mg^{2+} may increase the water's hardness, and this may result in the turbidity and precipitation of the juice. Raw materials that are too ripe, an excessive heating treatment, soft and rotten contents in the can, drastic shaking of the canned products during transportation, and flesh-scattered debris could also cause the turbidity and precipitation of the juice in the cans. Frozen canned foods may become mushy and fragmented, and bacteria could decompose the frozen food when it is thawed. Furthermore, the turbidity of the canned citrus fruits could result from the deposition of hesperidin. During the storage and sale process, hesperidin can be deposited from the citrus segments to form white crystals. These crystals can cause white patches on the citrus segments, turbidity of the juice, and affect the quality of the products. The aforementioned causes show that measures should be taken to avoid turbidity and precipitation.

2.6 NEW STERILIZATION TECHNIQUES OF CANNED FOODS

Canned foods are one of the most important Chinese exports. New technology was applied to the processing of canned foods, and this promoted the rapid development of the domestic canning industry. Although some techniques are not fully developed and limited in their practical use, their successful development is important to the canning industry. All of these novel techniques simplify the manufacturing procedure, enhance the production efficiency, and provide the benefits of nutrient preservation.

2.6.1 Microwave Technique for Food Sterilization and Preservation

The microwave was shown to have pyrolysis effects on electrolytes. Thus, the microwave has been widely used in the food industry to steam, boil, dry, bake, disinfect, and sterilize food products.

The microwave is a high-frequency electromagnetic wave (30–300 MHz) with polarity. Thus, it changes direction 2.45 billion times each second and is produced by magnetrons. When food is subjected to the microwave, the food's exterior and interior are simultaneously heated by absorption of the microwave's energy. The microorganisms' molecules in food are polarized to produce a high-frequency oscillation and thermal effect in the presence of the microwave field, which quickly increases the temperature. This rapid increase in temperature will change the protein structure and cause the loss of biological functions, which interrupts cell proliferation and cell death.

Microwave sterilization is a promising technique and has several advantages over the conventional sterilization methods. First, microwave sterilization is suitable for sterilizing foods before and after packaging. Microwave sterilization can be used for products that have plastic or composite films as the packaging material. If microwave sterilization is applied to packaged cans, the sterilization process must be conducted in a pressurized environment such as a pressurized glass container. Heating sterilization can produce large amounts of vapor, and this vapor creates high pressure inside the packages causing the packages to burst.

The microwave sterilization technique can also preserve food by deactivating the enzymes. Boiling is commonly used to remove most of the bacteria and deactivate the enzymes. The microwave sterilization technique does not cause the loss of water-soluble nutritional components such as vitamins, but the conventional method does. Thus, the microwave sterilization technique is a novel technique that is valuable to the canning industry.

2.6.2 Ohmic Sterilization

Sterilization of particulate foods (particle diameters of smaller than 15 mm) uses pipe and scraped surface heat exchangers for heat exchange. The heat exchange rate depends on the heat exchange conditions of conduction, convection, and radiation. During partition heat exchange, heat from a preheated medium such as steam is transferred through partitions to the liquid in the cans. In addition, heat is transferred from the liquid to solid particles through conduction and convection, and the heat is transferred internally in the solid particles to reach the sterilization temperature. When the temperature inside the solid particles reaches the sterilization temperature, the liquid is already overheated and will cause the food particles to become mushy and deformed, thus affecting the quality of the canned foods.

Ohmic sterilization is a novel technique that passes an electric current through the foods to achieve sterilization. This method has developed rapidly because of its quick temperature increase, uniform heating, no pollution, convenient operation, high thermal efficiency, and high product quality. The Ohmic heating method causes the heat-transferring rate in solid particles to be close to the heat-transferring rate in liquids. Ohmic heating can achieve a faster heating rate (1–2 °C/s) than the conventional heating method, which indicates that Ohmic heating shortens heating time and improves product quality. Furthermore, Ohmic heating does very little damage to the nutritional components and can be applied to foods with high density or high viscosity. The retention time of the sterilization temperature is largely reduced by Ohmic heating. The Ohmic heating principles focuses on quickly killing the bacteria by electrical heating and rupturing the cell membrane with the application of the electric field. This method removes *E. coli*, yeast, bacilli, and high resistance microorganisms. In these situations, the Ohmic heating device is placed in the pressured inert gas to improve sterilization efficiency.

Recently, APV Baker, a British company, developed Ohmic heating equipment for industrial use, which allowed the high-temperature short-time technique to be used for particulate foods (particle diameter is up to 25 mm). Since 1991, the Ohmic heating technique and equipment have been utilized in the production of low- and high-acidic foods.

2.6.3 Pulsed Light Sterilization Technology

Pulsed light sterilization is used to remove microorganisms with high-intensity light energy in a short amount of time. Electric energy is accumulated by the high-capacity capacitors, converted into high-density light energy and released in a short period. American researchers invented this technique, and this novel method does not impair the quality of foods compared with the conventional methods. Furthermore, foods can be sterilized after packaging because the pulse light has high energy. Pulsed light sterilization was largely unused in the food industry until a collaborative study on its application was completed by two Japanese companies (Nippon Meat Packers and Ishikawajima-Harima Heavy Industries Co., Ltd). These companies discovered that pulsed light sterilization is effective for all types of microorganisms, but bacilli show a higher resistance to the pulsed light as compared to the trophic bacteria. Sterilization efficiency is reduced when light with a wavelength of less than 300 nm is removed. This suggests that the pulsed light sterilization method is highly dependent on ultraviolet (UV) light. However, in dry conditions, the sterilization efficiency is retained after the light with a wavelength of less than 300 nm is removed. This suggests that pulsed light sterilization is greatly dependent on UV sterilization. Pulsed light sterilization has a shorter sterilization time than UV sterilization; however, pulsed light sterilization has certain disadvantages such as high radiation energy. To develop a device that converts the pulsed light sterilization parameters, the efficiency of pulsed light sterilization should be analyzed to reveal the correlation with the amplitude and no correlation with maximum energy. Therefore, maintaining the maximum energy and shortening the pulsed light amplitude can successfully maintain the sterilization efficiency and reduce the energy consumption by 30%.

Pulsed light treatment causes less damage to the bacterial plasmid DNA and yeast genomic DNA and causes more damage to the cell membrane of yeast than UV irradiation. Thus, pulsed light has sterilization mechanisms similar to those of UV. When the surface is irradiated by pulsed light, it will produce a metal-like odor, which shows that pulsed light sterilization and UV irradiation are similar and can denature or alter the components of the food. Additional studies have confirmed that the abnormal smell is still emitted from the denatured proteins irradiated by UV light with short wavelengths.

2.6.4 Ozone Sterilization Technique

Ozone is an allotropic form of oxygen and comprises three oxygen atoms. Ozone oxidizes microorganisms in air or water directly or indirectly, disrupting and degrading their cell walls, diffusing in the cells immediately, and thereby killing the pathogens. Ozone has an extensive elimination capacity to microorganisms, which reveals a sterilization efficiency 300- to 600-fold higher than that of chloride. At room temperature, ozone is decomposed into oxygen and active oxygen with a strong oxidation capability and a half-life of 12–16 h. The final product during ozone sterilization is oxygen; thus, the pollution caused by the residual ozone during the application of this technique is not a concern.

Ozone belongs to a gaseous sterilizing agent and ozone sterilization exerts comprehensive physical, chemical, and biological effects. The detailed mechanisms are discussed below:

(1) Ozone can increase the permeability of the cell membrane, cause the efflux of intracellular substances, and result in the loss of cell viability.

(2) Ozone can deactivate enzymes that are required for cellular activities. These enzymes are involved in basal metabolism and in the synthesis of some important cellular components.

(3) Ozone can impair genetic materials in cells leading to the loss of cellular functions. Ozone is generally speculated to damage viral DNA or RNA to achieve its sterilization role. In bacteria such as fungi, ozone first damages the cell membrane, leading to a metabolic disorder and then inhibition of bacterial growth. The continuous penetration of ozone in the cell membrane will result in its destruction (i.e., until the cell dies).

At present, a corona discharge ozone generator produces ozone. The glass tubes of the corona discharging equipment require relatively high intensity and smooth surface; however, in China, these requirements are currently not available. In China, it is almost impossible to produce glass tubes with relatively high hardness and strength. The ozone technology in China could not be well applied without the expansion of these factories which could produce such high-quality machineries. In the food industry, ozone sterilization disinfects by killing microorganisms, deodorizes by decomposing inorganic and organic compounds to remove odor pollution, and preserves fruits and vegetables by decomposing metabolic products and inhibiting the postripening process.

The application of the ozone technique in the food industry has a short history. By the end of 1908, the ozone water treatment was initiated in the Fuzhou water plant, which used an ozone generator from the German Siemens Company. In 1964, the Huangpu cold storage plant in the Guangdong province purchased an ozone generator with a production capacity of 5 g/h from the DEMAC Company to deodorize the freezer. Since 1986, ozone generators have been employed in dozens of large- and medium-sized businesses that store eggs, and excellent effects have been reported. At present, ozone generators are widely used in numerous fields such as the sterilization of mineral spring water, disinfection of uniforms and workshops, and preservation of meat, vegetables, and fruits.

2.6.5 High-Pressure Sterilization

High-pressure processing (HPP) sterilizes the sealed canned foods in a liquid medium or the liquid food is pumped into the treatment trough in the presence of 100–1000 MPa pressure. As early as 1899, American chemist Dr Bert Bite discovered that the storage life of milk was extended when the milk was subjected to a pressure treatment at 450 MPa. Thereafter, American physicist Dr Percy Williams Bridgman reported that hydrostatic pressure treatment (500 MPa) caused protein coagulation and that proteins could form gelatin under hydrostatic pressure treatment (700 MPa). However, at that time, these discoveries did not attract attention from the food industry. Professor Lin Lifan at Kyoto University in Japan investigated the use of high pressure in the food industry; since then, the studies on food processing with high pressure have increased. The mechanism of HPP is attributed to the destruction of protein conformation, protein denaturation, and even cell death. Common microorganisms perish with less than 450 MPa at room temperature but the elimination of spores requires a higher-pressure treatment or other treatments. Temperature increase is proportional to the pressure. The temperature of material increases by 2.4 °C per 100 MPa; thus, the lethal effect on microorganisms is due to the dual effects from heat and high pressure. In 1991, the first Japanese fruit jam was produced, followed by a high-pressure-treated juice. However, it was difficult to inactivate pectin methyl esterase (PME) in orange juice, which caused unstable precipitation of the juice. High-voltage electric field sterilization can be divided into batch and continuous sterilization.

Batch high-pressure processing (BHPP): This processing is similar to a batch sterilization machine. The juice is filled into the packages, and the packages are placed into high-pressure containers for sterilization. Sizer and his colleagues have designed high-pressure sterilization instruments. The entire sterilization procedure takes 5 min if these instruments are used. Sterilization requires 1 min for loading, 1 min for pressurization, 2 min for high-pressure processing, and 1 min for the pressure to be released.

Continuous high-pressure processing (CHPP): The products are directly pumped into a high-pressure container that has a baffle plate to separate the presser and juice products. Pressure is transferred to the products through the baffle plate. The pressure is released after treatment, and the products are pumped into sterilized cans. To prevent contamination, sterilized water is often used as the press medium. The advantage of the CHPP system is that HPP is connected with an aseptic packaging system to achieve continuous processing.

2.6.6 Pulsed Electric Field

Pulsed electric field (PEF) is another technique applied to food sterilization and storage.

Food is placed in a processing container with two electrodes, and an instant high-voltage electrical pulse is applied to the electrodes. Food in the container could be stored in long period due to the complete inactivation of microorganisms by PEF. The strength of the electric field in PEF is 15.80 kV/em, and the sterilization time is less than 1 s. Normally, the procedure can be completed within 100 μs. In the 1960s, the USA began to investigate the effects of PEF in food sterilization. China also started relevant research in the mid-1990s, when the technique was already in pilot scale production in the USA. PEF has two mechanisms in sterilization: electroporation and electrical breakdown. Electroporation shrinks the cell membrane and forms pores in the applied electronic field, leading to an increase in the permeability of the cell membrane. Small molecules such as water penetrate the cell membrane and move into the cytosol, which results in cell expansion and bursting of the cell membrane. Cell death is due to the leakage of the cytosol. In the electric breakdown theory, the cell membrane of the microorganisms is considered an electric container filled with electrolytes. The potential difference (V) of the cell membrane is formed when an electric charge is distributed and separated across the membrane under an external electric field. V is proportional to the intensity of the electric field and the diameter of the cells. Once the intensity of the external electronic field is enhanced, the membrane V increases and the cell membranes become thinner. When the membrane V reaches the threshold for the difference, breakdown (K) occurs and the cell membrane ruptures. The pores on the cell membrane are formed (with electrolytes), followed by an instant discharge on the cell membrane and bursting of the membrane. If the pores occupy a small portion of the cell membrane, then the breakdown of the cell membrane is reversible. If the cell membrane is maintained in an intense electric field (i.e., above the threshold for a long period), this condition will cause a large area of the cell membrane to rupture. The breakdown of the cell membrane will be irreversible, leading to the death of the microorganisms (Figure 2.4).

The PEF technique is mainly used in the sterilization of fluid foods. Reports have shown that the PEF is very effective in sterilizing juice. Compared with conventional pasteurization, PEF sterilization is superior in saving energy, reducing cost, and making adequate profits. Figure 2.5 shows the PEF sterilization flowchart.

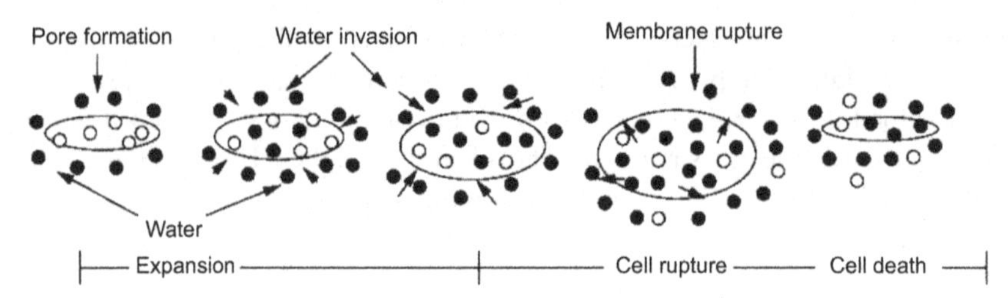

FIGURE 2.4 Procedure of cell perforation.

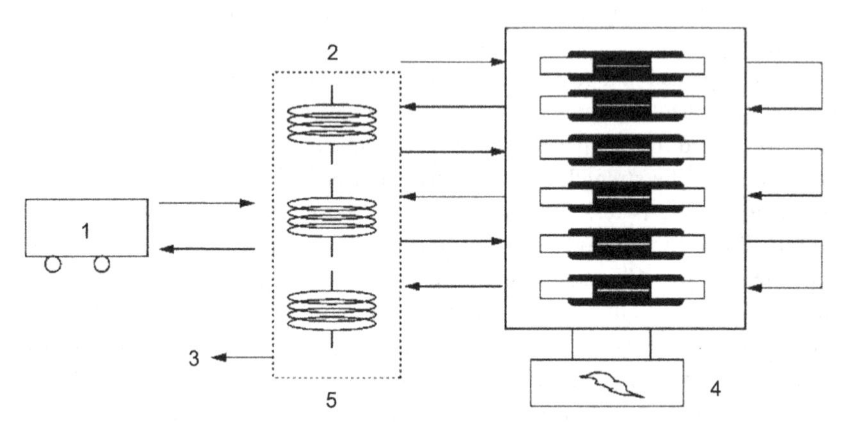

FIGURE 2.5 Diagram of PEF procedure system. 1. Condenser; 2. Sample inlet; 3. Sample outlet; 4. Pulse generator; 5. Temperature control.

2.6.7 Electrostatic Sterilization

Developed countries have already grasped the importance of electrostatic technology in food sterilization. The mechanism uses an ion atmosphere or ozone, which is generated by an electrostatic field to sterilize foods. This technique can achieve excellent sterilization and preservation effects. Ozone can kill bacteria and molds in grain, fruits, bottles, tanks, and bags, as well those in the storage room. The sterilization speed is approximately 15–30 times faster than oxygen. Furthermore, when the components of food were subjected to electrostatic sterilization, no changes were observed in food when it was compared to food without sterilization treatment. Thus, this technique protects the original flavors of food.

2.6.8 Magnetic Sterilization

In Japan, Akita University and the Akita Brewage Institute collaborated and succeeded in constructing variable magnetic sterilization. Magnetic sterilization uses a magnetic field (up to 600 G), and food is placed between two poles. After continuous shaking, sterilization can be completed without heating. This technique maintains the original components and flavors of food.

2.6.9 Induction Electronic Sterilization

Ionized radiation is generated by a linear induction electron accelerator that uses electricity as its energy source and causes chemical reactions in microorganisms and food to deactivate and eliminate microorganisms. This advanced technique accelerates electrons to strike the lead plate, which causes the plate to emit a strong radiation that has a long wavelength spectrum. Thus, this convenient technique uses high energy to sterilize food.

2.6.10 Other Sterilization Techniques

(1) X-ray sterilization: The Toyo Automatic Machinery Company and Tengsen Industrial Company developed a chain-belt style aseptic filling and food packaging machine that uses X-ray sterilization. This machine uses X-rays to presterilize the food. It also uses bags that are prepared and individually sent to the packaging machine for sterilization. Other machines that use a similar sterilization process (i.e., individual bags are sent to the packaging machine for sterilization) differ from this machine because this machine uses composite films as raw materials to prepare the package bags. Moreover, the sterilized bags sent to the filling and packaging machine are sealed, which simplifies the process of filling and heat sterilization. This X-ray sterilization technique is fast and does not damage the flavor components of the food.

(2) Infrared sterilization: The Japanese Sanzhi Company pioneered the development of the aseptic packaging machines. The entire machine is composed of a packaging machine (ML-501) and a channel infrared heat shrinking machine (MS-801). This machine selects heat shrink films with an appropriate thickness and color based on the size and shape of the food and uses infrared light to sterilize the food. This technique increases the packaging rate by six- to eightfold due to its simplicity.

(3) Nuclear radiation sterilization: This method was developed by the American food industry. Nuclear radiation sterilization uses an ionization ray (usually γ-rays) emitted by dual radiation elements during decay. ^{60}Co or ^{137}Cs is usually the radiation source. The γ-rays, with their extremely short wavelengths, can penetrate solid substances and impair the cell walls of all microorganisms and viruses to achieve sterilization. This is known as the "cold treatment," which indicates that no temperature increase is observed during the treatment.

(4) Chitin sterilization: Chitosan is an antimicrobial agent that has amazing properties against fungi. American researchers have reported that chitosan inhibits the growth of *Saccharomyces cerevisiae* in the fermentation of bread and in the brewing of wine. High-pressure treatment or high-pressure homogenization enhances the sterilization efficiency of chitosan. The antimicrobial mechanism of chitosan is due its ability to form a cation polymer coating on the surface of the microorganisms that prevents them from communicating with their environment.

(5) Sterilization by antimicrobial enzymes: In Japan and the USA, this food sterilization technique is gaining tremendous attention. Antimicrobial enzymes such as chitinase and glucanase inhibit the growth of Gram-positive bacteria by impairing their cell membrane. At present, antimicrobial enzymes inhibit cell metabolism, produce toxic substances, damage the cell membrane, and deactivate cellular enzymes. Enzymes that inhibit cell metabolism, produce toxic substances, and damage the cell membrane have already been employed in the food industry.

Chapter | Three

Machinery and Equipment for Canned Citrus Product Processing

CHAPTER OUTLINE

3.1 Material Handling Machinery and Equipment....48
- 3.1.1 Belt-Driven Conveyors....48
- 3.1.2 Bucket Elevator....49
- 3.1.3 Spiral Conveyor....51

3.2 Cleaning Machinery and Equipment....51
- 3.2.1 Cleaning Machinery for Raw Citrus Materials....52
- 3.2.2 Cleaning Machinery for Packaging Containers....54

3.3 Machinery and Equipment for Processing Raw Citrus Materials and Semifinished Products....59
- 3.3.1 Citrus-Sorting Machine....59
- 3.3.2 Scraper-Style Continuous Citrus Peel–Heating Machine....72
- 3.3.3 Orange Segment–Sorting Machine....73
- 3.3.4 Pilot Equipment for Removing Citrus Sac Coating by Complex Enzyme....73

3.4 Exhaust and Sterilization Machinery and Equipment....76
- 3.4.1 Exhaust Machinery and Equipment....76
- 3.4.2 Sterilization Equipment....79

3.5 Packaging Machinery and Equipment....86
- 3.5.1 GT786 Automatic Vacuum Juice-Filling Machine....86
- 3.5.2 Sealing Machinery for Canned Citrus Segment Products....87
- 3.5.3 Spinning Capper....91
- 3.5.4 Vacuum-Packaging Machine....94
- 3.5.5 Labeling Machine....96
- 3.5.6 Packing Machine....98
- 3.5.7 Carton-Sealing Machine....101
- 3.5.8 Strapping Machine....102

3.6 Typical Canned Citrus–Processing Production Line....104

Y. Shan (Ed): Canned Citrus Processing. http://dx.doi.org/10.1016/B978-0-12-804701-9.00003-4

3.1 MATERIAL HANDLING MACHINERY AND EQUIPMENT

During canned product processing, various transport machines are involved in the delivery of raw materials, including all of the processes from the raw materials to the final products, and each process from one manufacturing unit to a subsequent manufacturing unit. In particular, when advanced technology and equipment are applied, including the automation of single equipment units, systematic connections among these single manufacturing units are essential to establish an automatic production line. Moreover, under large-scale production conditions, transfer machines and equipment are necessary to improve productivity.

3.1.1 Belt-Driven Conveyors

Belt-driven conveyors are widely used in citrus-processing factories as continuous conveyor machines. They are suitable for citrus cleaning, selection, handling, and raw material inspection. They are mainly used in the raw material pretreatment processes, product delivery, material selection and filling, packaging, and warehouse storage.

The conveyor speed selection depends on the specific process requirements. When the conveyor is used as a transporter, its speed ranges from 0.8 to 2.5 m/s. When it is used as a checkpoint, its speed should be controlled to 0.05–0.1 m/s.

A belt-driven conveyor is shown in Figure 3.1, which represents flexible driving equipment.

During the operation of the driving unit, driving roller 8 rotates in a clockwise direction. Belt 6 moves forward because of the friction between driving roller 8 and the inner surface of belt 6. After the start phase has stabilized, raw materials are fed through loading hopper 3 and moved forward by closed loop belt 6 to the working location. If the direction of transportation needs to be changed, unloading device 7 will rotate to transfer the raw materials to another conveyor. Otherwise, unloading device 7 remains unchanged until the raw materials are discharged from the correct end.

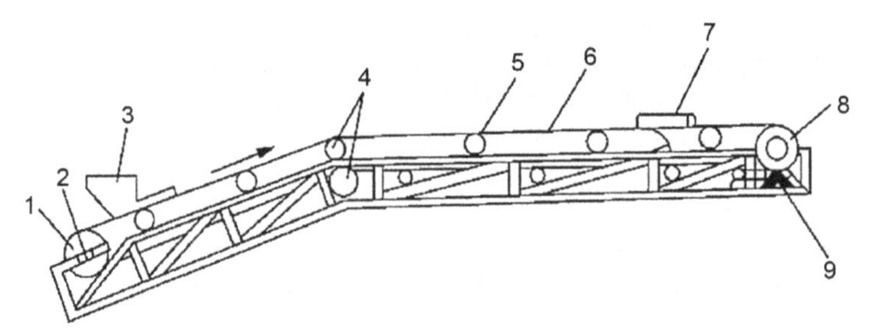

FIGURE 3.1 Belt-driven conveyor. 1. Tension roller; 2. tension unit; 3. loading hopper; 4. reorientation roller; 5. supporting roller; 6. closed loop belt; 7. unloading device; 8. driving roller; 9. driving unit.

3.1.2 Bucket Elevator

During the continuous manufacturing of canned citrus products, the raw materials sometimes need to be transported at different heights. For example, the juice pressing process and canned product manufacturing process operate at different heights. Thus, the materials need to be elevated in a vertical direction or almost perpendicular to the conveyor movement direction. In these cases, a bucket elevator is often used.

Bucket elevators are mainly used for the transportation of raw materials among different heights. They are suitable for elevating raw citrus materials from a low position to a higher position. The major advantages of bucket elevators are small space, large lifting height (typically 7–10 m, maximum height up to 30–50 m), and high productivity (3–160 m^3/h).

Bucket elevators can be divided into tilt elevators and vertical elevators according to their transportation direction.

Bucket elevator mainly comprise elevator tractors, rollers (or chains), tension devices, loading and unloading devices, driving devices, and hoppers. A series of small buckets (called hoppers) are installed on the elevator tractor, which move upward with the tractor and flip over at the top to discharge the raw materials. The back of the bucket (posterior wall) is often mounted on a leash or chain. In double-chain elevators, the bucket sides are sometimes fixed to chains.

A tilt bucket elevator is shown in Figure 3.2. The lift height is adjustable to meet the requirements of different production conditions. There is a removable section in the bucket slot, which can be used to extend or shorten the lifting distance of the bucket elevator. The brackets are also retractable and

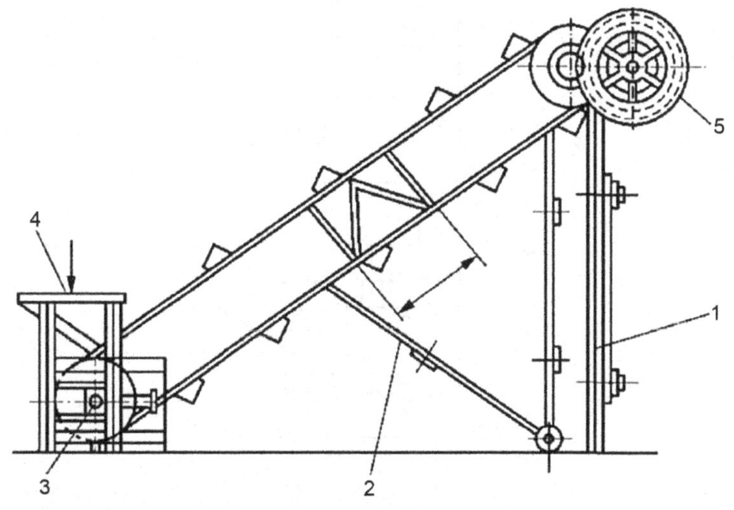

FIGURE 3.2 Tilt bucket elevator. l, 2. Brackets; 3. tension device; 4. material feeding port; 5. driving unit.

are often fixed with screws. The brackets are either vertical (e.g., bracket 1) or inclined (e.g., bracket 2). An inclined bracket is fixed in the middle of the bucket slot. To move more easily, the brackets are sometimes mounted on movable wheels.

A vertical bucket elevator is shown in Figure 3.3, which consists of a hopper, traction belt (or chain), driving unit, chassis, and loading and uploading ports. When the raw materials are fed uniformly through the loading port, the hopper fixed on the conveyor can scrape the raw material under the driving of the driving roller. The scraped raw materials move upward with the conveyor and flip over at the top to discharge the raw materials, by either centrifugal force or gravity.

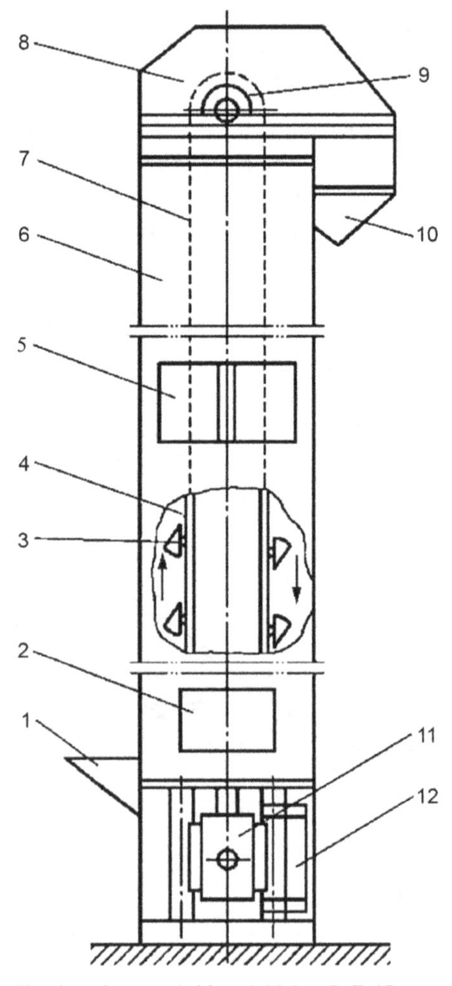

FIGURE 3.3 Vertical bucket elevator. 1. Material inlet; 2, 5, 12. openings; 3. hopper; 4, 7. conveyor; 6. shell; 8. driving drum shell; 9. driving drum; 10. discharging port; 11. tension device.

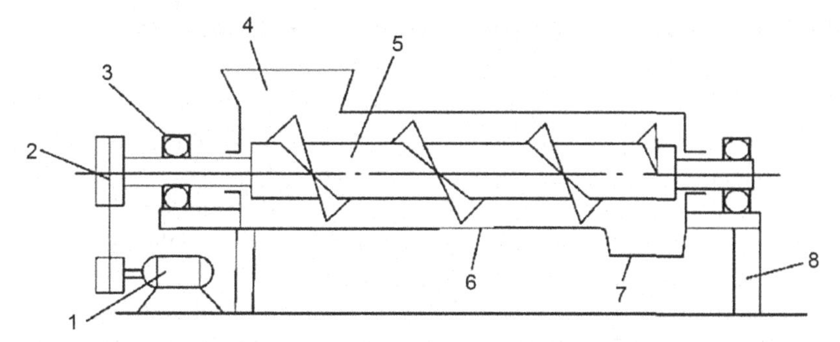

FIGURE 3.4 Schematic diagram of a spiral transporter. 1. Motor; 2. driving unit; 3. bearing; 4. material inlet; 5. conveyor spiral; 6. material chute; 7. discharging port; 8. mechanical frame.

3.1.3 Spiral Conveyor

A spiral conveyor is commonly known as an "auger," which uses spiral rotation to move raw materials in a forward direction. This type of machine is mainly used for the short-distance transportation of small amount of materials, such as citrus peel.

The schematic structure of a spiral conveyor is shown in Figure 3.4, which consists of the motor, transmission unit, bearing, loading hopper, spiral conveyor, material chute, discharging port, and mechanical frame. The key components of the spiral conveyor are the helical blade shaft (often called a conveyor spiral) and the material chute (shell). The loading hopper is located at the top of the material chute and the discharging port is located at the bottom of the material chute.

During operation, the driving force is provided by the transmission unit and the spiral conveyor begins to rotate. After the start phase has stabilized, raw materials are loaded from the loading hopper and pushed forward along the material chute via the axial thrust generated by the rotation of the spiral conveyor. There is sufficient frictional resistance between the material and the wall of the material chute. Thus, the raw materials will not rotate together with the helical blade shaft. Therefore, the raw materials move forward between the blades along the axis of the material chute and are finally discharged through the discharging port.

3.2 CLEANING MACHINERY AND EQUIPMENT

Citrus materials are vulnerable to dust, microbes, and other dirt contamination during the growth, maturation, transportation, storage, and manufacturing processes. Therefore, citrus materials must be cleaned before processing. In addition, to ensure the quality of the final products, appropriate cleaning machinery and equipment are required in the manufacturing processes, including equipment for canned product storage, connecting pipelines, and packaging containers.

3.2.1 Cleaning Machinery for Raw Citrus Materials

1. GT5A9 fruit-brushing machine

The GT5A9 fruit-brushing machine is shown in Figure 3.5, which consists of the loading and uploading hoppers, vertical and horizontal brushing rollers, transmission system, chassis, and other components. Motor 6 drives the vertical brushing rollers, which spin through the intermediate shaft and sprocket via the V-belt to drive the horizontal brushing roller that spins through the chain wheel of the vertical brushing rollers, cone gear, and V-belt.

Citrus fruits are loaded from the hopper into the vertical brushing rollers. The tufts on the brushing roller are arranged spirally and grouped alternately in short and long groups. Two adjacent brushing rollers rotate in an opposite direction, and the axis of the roller has a 3°–5° inclination relative to the horizontal axis. The fruits are rotated and pushed forward while being brushed, and they finally reach the horizontal brushing rollers. The horizontal brushing rollers brush and clean the fruits, but they also control the discharging speed so that the fruits can be cleaned better.

2. DT5A1 fruit-washing machine

The DT5A1 fruit-washing machine is mainly used to wash raw fruits, and it is widely applied in juice production lines. It consists of rolling conveyor 6 and washing tub 3, as shown in Figure 3.6. The roller conveyor is similar to the belt conveyor, except that several 76-mm cylindrical rollers are installed between two chains, where the space between the rollers is 10mm. When the chains are driven to move, the raw materials move along the rollers. The conveyor is divided into three sections. The inclined lower section is immersed in washing tank 3. The inclined upper section is connected with the crusher. The intermediate horizontal section is used for inspection. There are four water-spraying pipes in each inclined section. Each water-spraying pipe has two rows of holes with a perpendicular

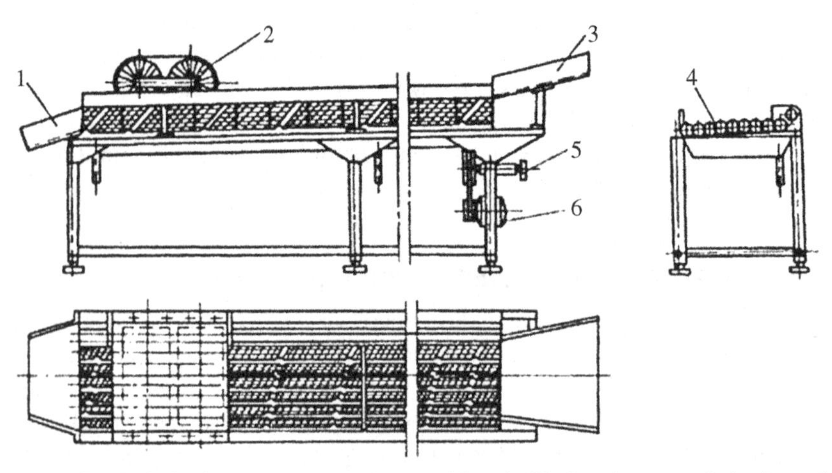

FIGURE 3.5 GT5A9-type fruit-brushing machine. 1. Discharging port; 2. horizontal brushing roller; 3. material feeding hopper; 4. vertical brushing roller; 5. transmission system; 6. motor.

arrangement. The DT5A1 fruit-washing machine is typically equipped with conveyors to transport raw materials.

During operation, the raw materials are prewashed and transported by the conveyor, before they are elevated by lifting unit 1 and immersed in the front of washing tank 3. Next, wheel 2 turns the fruits to the back of the washing tank, where high-pressure water hose 7 with holes at the same distance is installed. High-pressure water is ejected from the holes and the fruits are flipped, which generates friction among the fruits, thereby removing dirt from the surfaces of the fruits. The fruits are washed again by high-pressure water pipes 4 on conveyor 6. The fruits then proceed to inspection station 5, where rotten and defective fruits are removed. After spraying, the fruits proceed to the next operation. This machine has a productivity of 10 t/h, rotation speed of 31 r/min, and dimensions of 14,540 × 2700 × 2600 mm.

3. Drum-type fruit-washing machine

The drum-type fruit-washing machine uses the rotation of the drum to keep the fruits tumbling to facilitate fruit washing. High-pressure water is also sprayed to wash the fruits. Dirty water and sediments are discharged through the drum mesh via the bottom hopper. This machine is suitable for cleaning fresh fruits such as citrus fruits.

A continuously operating drum-type fruit-washing machine is shown in Figure 3.7. Drum 3 is welded using angled steel, flat steel, and steel plates.

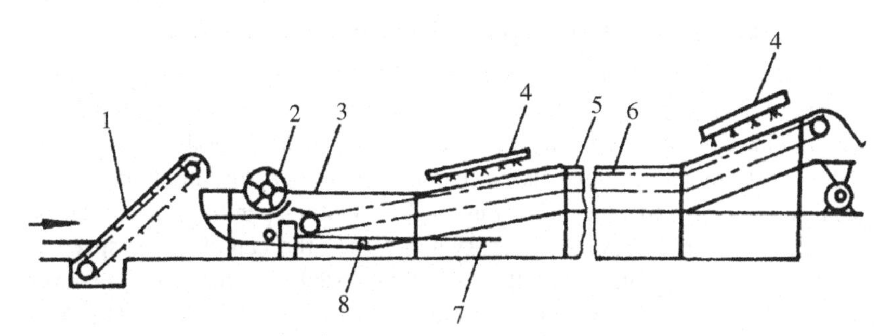

FIGURE 3.6 Fruit-washing machine. 1. Lifting unit; 2. fruit-flipping wheel; 3. washing tank; 4. water-spraying pipes; 5. selection station; 6. rolling conveyor; 7. high-pressure hose; 8. drainage.

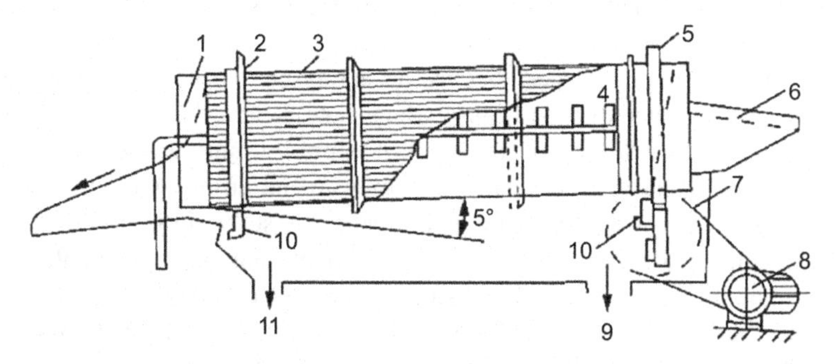

FIGURE 3.7 Drum-type fruit-washing machine. 1. Raw material outlet; 2. roller bracket; 3. drum; 4. nozzle; 5. drum-supporting roller; 6. raw material inlet; 7. gear; 8. motor; 9. fluid exit; 10. drum bracket; 11. catchment outlet.

The drum axis has an inclination of 3°–5°. The spiral guiding plates are installed on the inner wall of the drum to facilitate material discharge. The raw materials are fed through inlet 6, and they tumble during the drum rotation process. The friction between the raw materials, raw material and water, and the raw material and the inner surface of the drum can remove the dirt. High-pressure water is also sprayed continuously from nozzle 4 to immerse the raw materials and remove the dirt with water. The cleaned raw materials are discharged from outlet 1.

To improve the cleaning efficiency, some drum-style washing machines also have vertical and horizontal adjustable brushes. These brushes can enhance the removal of dirt from citrus fruits. The washing rate of the drum-type machine is up to 99%, and its production capacity is up to 1000 kg/h.

The drum-style fruit-washing machine is generally driven by supporting wheels and rolling rings. There are three to four rolling rings, which are fixed outside the drum. Each ring is supported by supporting wheels on both sides. Motor 8 drives the supporting wheels via transmission device 7, and the supporting wheels drive the drum to rotate through the rolling rings via friction. This driving style has the advantages of a simple structure and stable transmission; thus, it is applied widely in industrial production lines.

3.2.2 Cleaning Machinery for Packaging Containers

1. Three-piece can-cleaning machine

Figure 3.8 shows the widely used three-piece can-cleaning machine. The machine has a box-type structure, which mainly consists of the rack cabinet, loading and uploading round discs, magnetic conveyor, transmission system, cleaning nozzles, and other components. The motor operating speed can be adjusted using an electronic speed switch. Thus, the can-cleaning speed is adjustable, and it should match the can-filling speed. The magnetic conveyor consists of nylon belt 8, which is tensioned by two magnetic drums 3. The magnetic conveyor is mounted on cabinet 7 and driven by the transmission system. Supporting discs 1 are fixed on both sides of the cabinet, which can be used to support the nylon belt. Magnetic plate 2 is the same width as the nylon belt and it is mounted on both sides of the cabinet. It runs parallel to the inner surface of the nylon belt.

When the cleaning process begins, an empty can is placed on can-feeding disc 6. Then, it moves via can-guiding plate 5 onto the conveyor. When it reaches magnetic drum 3 on the right, it is pressed onto the surface of the conveyor via magnetic force and it rotates with the drum. When the empty can moves to the lower part of the nylon belt, it is still held by the magnetic force (magnet plate 2) and moves with the belt from the right to the left. At the same time, the nozzles spray water to wash the can. As the can moves to the upper part of the drum on the left, the can-guiding plate and uploading disc move the can to the can-filling machine, where it awaits the filling process.

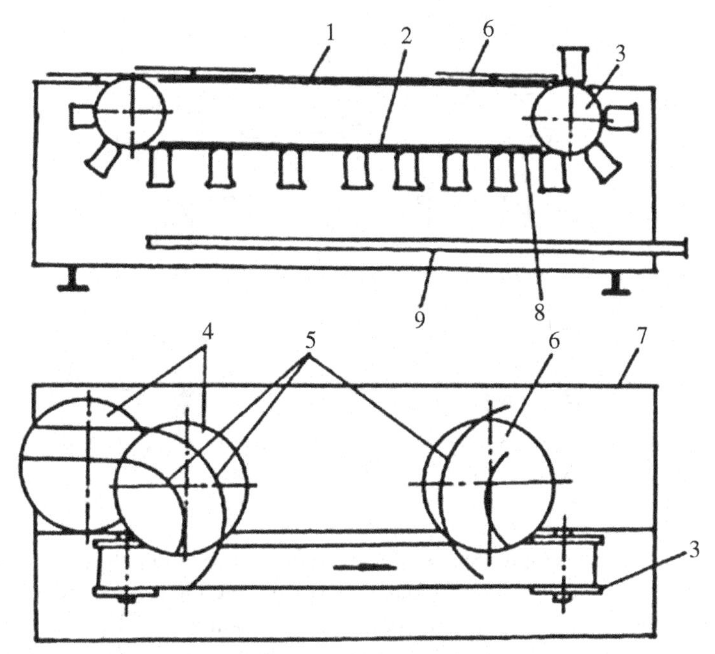

FIGURE 3.8 Three-piece can-cleaning machine. 1. Supporting disc; 2. magnetic plate; 3. magnetic drum; 4. tank-discharging disc; 5. can-guiding plate; 6. can-feeding disc; 7. rack cabinet; 8. nylon belt; 9. water-spraying pipe.

2. Cleaning machine for filled cans

After the cleaning processes, the insides and outsides of the empty cans are clean. However, after the filling, exhausting, sealing, and sterilization processes, the surface of the filled can is often coated with soup or the filling contents, which may become greasy or sticky with dark strips or spots after the sterilization process. In particular, cans that are damaged by the sterilization process may leak to release soup or other contents, thereby contaminating the surfaces of the cans. These dirty patches will become firm and stick tightly to the surface after heating and cooling processes. Therefore, surface cleaning is required before and after the sterilization processes.

As shown in Figure 3.9, the machine used to clean filled cans mainly consists of the lye pool 4, water tank 3, and dryer 2. T transmission system and conveyor link the three parts to form an overall system.

First, the filled can is immersed in the lye pool for cleaning, before cleaning in the fresh water tank. Finally, it is dried in a dryer. A spraying device is installed between the lye pool and the fresh water tank, which washes most of the lye liquid from the surfaces of the filled cans. To improve the drying efficacy and to reduce the drying time, a blowing device is installed between the lye pool and dryer, which blows some of the water from the surfaces of the filled cans.

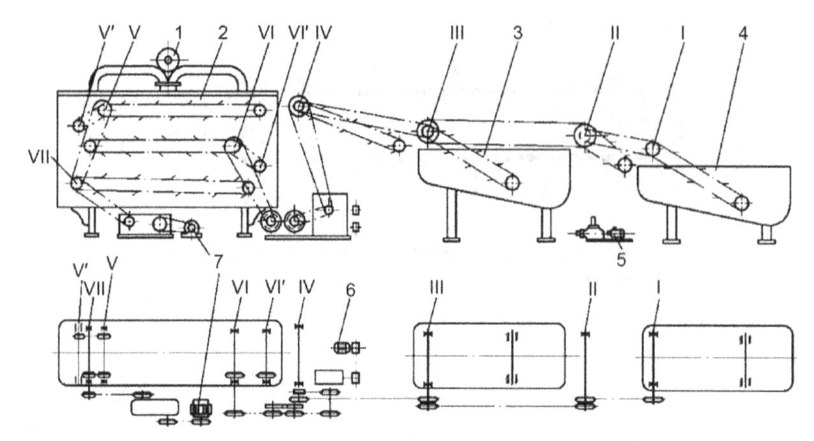

FIGURE 3.9 Machine for cleaning the surfaces of filled cans. 1, 5, 6, 7. Motors; 2. dryer; 3. fresh water tank; 4. lye pool; I, II, III, IV, V, VI, VII, V′, VI′ axes.

This machine has multiple advantages in terms of its strong decontamination capacity, good mechanical performance, large production capacity, convenient operation and management, compact structure, continuous cleaning process, flexibility for various can types, and low manufacturing cost. However, it also has the disadvantages of manual loading and uploading, large fluctuations in the lye solution and temperature due to the nonautomation of this process, and low drying efficiency.

There are four motors in the machine. Motor 1 is used by the dryer to provide the exhaust. Motor 5 is used to drive the water cycle to generate centrifugal force. Thus, only motor 6 and motor 7 operate the transmission system. These two motors allow the transmission system to work in two modes.

In the first mode, motor 6 drives the pulley and gearbox. The output axis of the gearbox has double sprocket wheels. One wheel connects the sprocket wheel to shaft IV via a chain. Shafts IV, III, and I are driven by chains. Thus, the lye washing, fresh water washing, transition, and blowing processes operate continuously. Using a pair of gears, another wheel transfers the power to shaft VI and then to shaft VI′. This mode drives the middle-layer conveyor system in the dryer and the movement of the dialing wheel from the middle to the lower level to allow can transfer.

In the second mode, motor seven transmits power via the continuously variable transmission (CVT) to shaft VII. Shaft VII drives shafts V and V′ through chains. This mode drives the upper and lower level conveyors in the dryer and the movement of the dialing wheel from the upper to the middle level to allow can transfers.

3. Automatic bottle-washing machine

To ensure quality products, packaging containers must be cleaned before filling with citrus juice. Medium- and large-scale citrus beverage-processing plants typically use automatic bottle-washing machines. Two types of bottle-washing machines are used widely: jetting-type and jet-soaking-type bottle-washing machines. The first type uses a high-pressure nozzle to clean the inner and outer surfaces of the bottles several times, including preheat spraying, lye spraying several times, and backward water spraying (hot, warm, and cold water), as well as a final spraying with fresh water. The second type uses continuous or intermittent soaking and multiple sprays for several cycles to obtain the best cleaning efficiency. In general, this type includes presoaking, multiple lye-soaking, lye-spraying, hot water-spraying, warm water-spraying, cold water-spraying, and fresh water-spraying processes.

1) Structure of the automatic bottle-washing machine

The automatic bottle-washing machine mainly consists of the box-type chassis, transmission system, several chain belts for bottle transfer, several driving units, tension and redirection chain wheels, presoaking tank, detergent-soaking tank, cooling water tank, disinfection tank, pump-spraying device, and loading and uploading devices.

A section of a double-ended bottle washer is shown in Figure 3.10. During operation, the bottles are loaded by the loading device from one end, and they move along the conveyor to undergo the entire washing process, before they are discharged at the other end by the unloading device.

Figure 3.11 shows a section of a single-ended bottle washer, where the cleaned bottles are discharged from the same end.

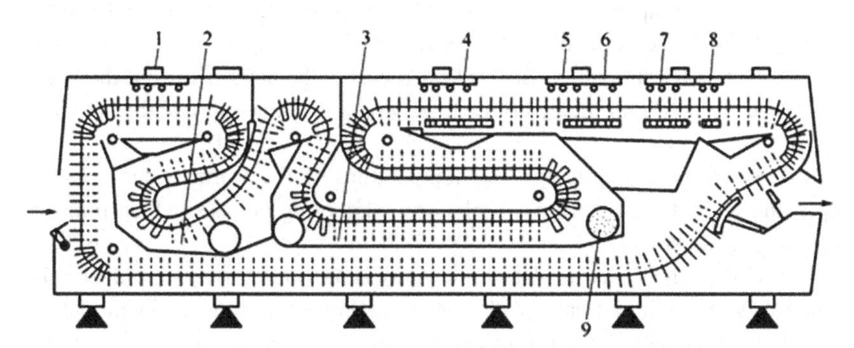

FIGURE 3.10 Double-ended bottle washer. 1. Prewashing nozzle; 2. presoaking tank; 3. detergent-soaking tank; 4. detergent nozzle; 5. hot water-prespraying nozzle; 6. hot water spraying; 7. warm water spraying; 8. chlorine spraying; 9. central heater.

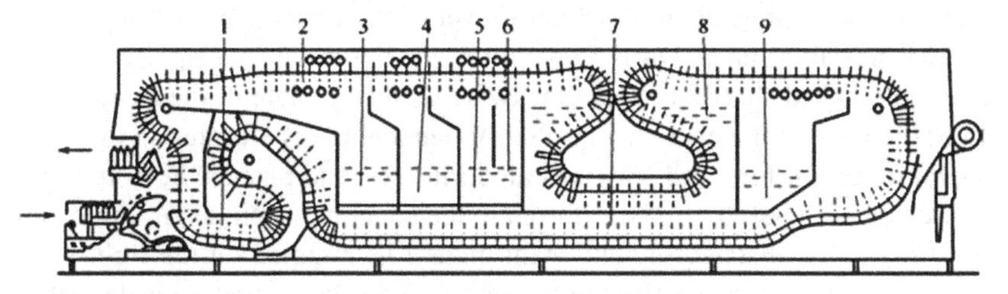

FIGURE 3.11 Single-ended bottle washer. 1. Presoaking tank; 2. chlorine spraying; 3. cold water spraying; 4. warm water spraying; 5, 6. hot water spraying; 7, 8. detergent soaking; 9. detergent spraying.

2) Cleaning process

The processes performed by the automatic bottle-washing machine are divided into six steps.

(1) Prewashing and presoaking: These processes remove loose debris from the bottle to minimize adsorption in the later soaking processes and to fill the bottle before preheating. To avoid broken bottles, the temperature difference between the solution and the bottle should not exceed 30 °C. The prewashing temperature is 30–40 °C.

(2) Detergent soaking: After the prewashing process, the bottles are transferred to the liquid detergent tank for soaking. This process aims to dissolve impurities and emulsified fats, thereby facilitating the removal of impurities. The washing efficiency depends on the soaking time and the solution temperature. In general, the detergent temperature is controlled to 65–70 °C, and the detergent concentration is 1–1.5%.

(3) Detergent spraying: When the conveyor transfers the bottles from the soaking tank to the detergent-spraying area, the dissolved contaminants are washed and removed by detergent spraying at a spraying pressure >0.2 MPa. The detergent-spraying temperature is 70 °C. During the detergent-spraying process, the attachment of foam to the bottles is detrimental. The major causes of foam formation are as follows: a weak water pump seal, which results in the entry of air; the detergent-spraying pressure is too high; and dirty bottles with oily residues. Pump maintenance and improved spraying systems are the major strategies used to eliminate foam formation.

(4) Hot water spraying: This process removes the detergent from the bottle and cools the bottle for the first time. In general, the temperature of the hot water is 55 °C.

(5) Warm water spraying: The temperature of the warm water used for spraying is 35 °C. This process is the second cooling process, which removes the residual detergent.

(6) Cold water spraying: This process cools the bottles to room temperature. The cold water must be treated with chlorine to prevent secondary contamination.

3) Bottle-loading and -unloading devices

Automatic bottle-loading and -unloading devices are shown in Figure 3.12(a) and (b).

The cam plate 4 is mounted on the shaft 2. On each side of the cam plate, there is a clamping plate 3, which prevents the bottle from tilting to the side as the cam rotates. During operation, the bottles are transported by the conveyor. When shaft 2 rotates clockwise with cam plate 4, the bottles that enter the neck of the cam are elevated. When a bottle is elevated in the horizontal place, pushing rod 5 swings along its axis and pushes the bottle into bottle holder 7, before it is transferred to the cleaning machine via chain belt 6. At the bottle-discharging end, cam plate 4 is mounted on the right-hand side of the conveyor. As the bottle moves along the conveyor, it describes arc supporting board 9 due to its own weight and the bottle drops to sliding

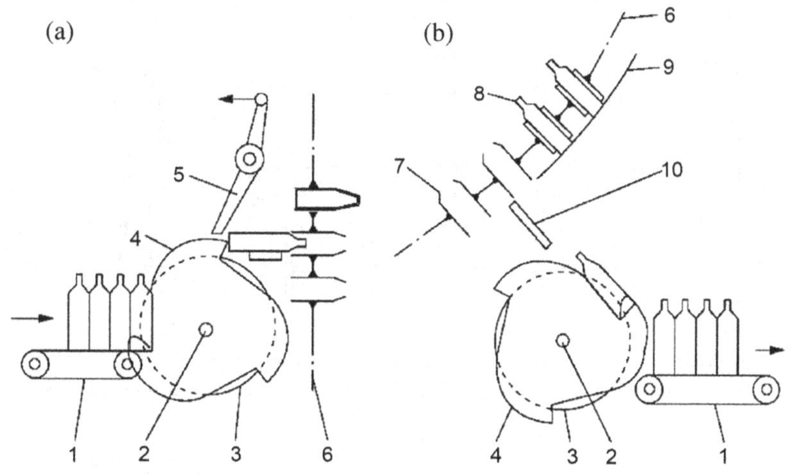

FIGURE 3.12 Bottle-loading and -unloading devices. (a) Bottle-loading device; (b) bottle-unloading device. 1. Conveyor; 2. cam shaft; 3. clamping plate; 4. cam plate; 5. bottle-feeding pushing rod; 6. chain belt; 7. bottle holder; 8. bottle; 9. arc supporting board; 10. sliding plate.

plate 10, before moving to the bottle reception part on cam plate 4 for discharging.

3.3 MACHINERY AND EQUIPMENT FOR PROCESSING RAW CITRUS MATERIALS AND SEMIFINISHED PRODUCTS

3.3.1 Citrus-Sorting Machine

1. Roller-style citrus sorter

The structure of the roller-style citrus sorter is shown in Figure 3.13. Its major operating principles are as follows. Raw citrus materials flow into the drum through the hopper and roll in the drum. During this process, the raw citrus materials pass through appropriate holes to complete the sorting procedure.

The roller-style citrus sorter is characterized by its simple structure, high sorting efficiency, smooth operation, and lack of unbalanced power. Major components of the roller-style citrus sorter are as follows:

(1) Roller: This is a drum with apertures, where the drum is divided into several segments (groups) according to the specific sorting requirements. Different segments have different apertures, but the apertures are of same size in each segment. The aperture in the inlet end is the smallest and the aperture in the outlet end is the biggest. Each segment has a funnel device that collects the sorted raw materials.

The raw materials fall from the inlet end and move forward in the drum while it rotates. Finally, the raw materials are discharged from the holes in the drum via each segment, according to their size.

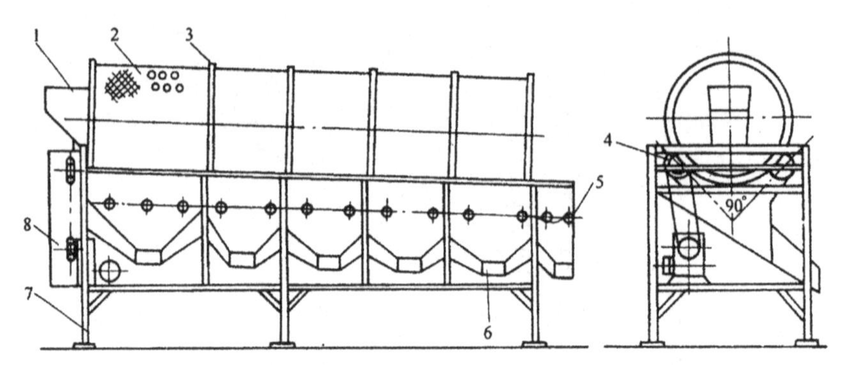

FIGURE 3.13 Roller-style citrus sorter. 1. Material-feeding hopper; 2. roller; 3. rolling ring; 4. friction wheel; 5. hinges; 6. collecting funnel; 7. mechanical rack; 8. transmission system.

(2) Supporting device: The supporting device consists of rolling ring 3, friction wheel 4, and mechanical rack 7. The rolling ring is mounted on the drum and it transfers the weight of the drum body to the friction wheel. The entire device is supported by the mechanical rack.

(3) Collecting hopper: The collecting hopper is located below the drum. The number of hoppers is the same as the number of sizes sorted.

(4) Transmission device: At present, the most widely used transmission device is friction wheel transmission.

(5) Sieve-cleaning device: During the operation process, raw citrus materials pass through the openings in the drum via the corresponding apertures to allow sorting. However, the openings can be blocked by raw materials, which will affect the sorting efficiency. Therefore, the sieve-cleaning device is essential for ensuring the discharge of raw materials from the corresponding apertures. The mechanical sieve-cleaning device is a wooden roller, which is installed on the outer wall of the drum. The wooden roller runs parallel to the central axis of the drum and it pinches the outer wall of the drum via a spring. Squeezing by the wooden roller allows the raw materials that block the openings in the drum to be pushed back into the drum.

Figure 3.14 shows another roller-type citrus-sorting machine that uses a hollow drum. The raw citrus materials are transported through the outer surface of each drum. Each drum has openings with apertures of SS, S, M, L, and LL. The drums are placed side by side. The raw materials enter from the top and they are sorted according to their size, from very small to very large. Depending on the factory size and the raw material volume, the optimal number of drums is 2–4. The sizes of the raw material components should be matched with the apertures.

2. Three-roller citrus sorter

This machine is suitable for sorting based on the size of citrus fruits, as shown in Figure 3.15. The overall machine consists of a material-feeding hopper 1, material-organizing drum 2, roller chain belt, material-discharging conveyor 7, hoistway 9, and driving device. The sorting component is a

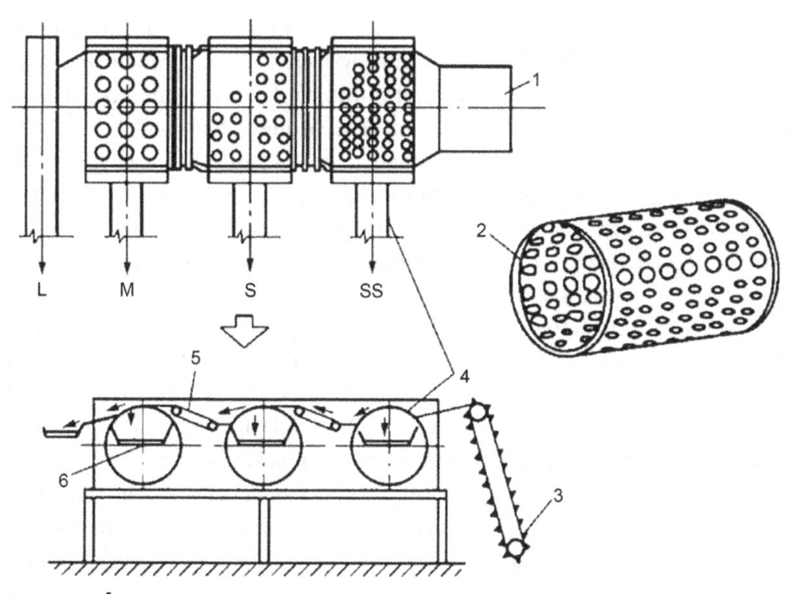

FIGURE 3.14 Drum-style citrus sorter. 1, 3. Raw material; 2. drum; 4. input conveyor; 5. roller conveyor; 6. output conveyor.

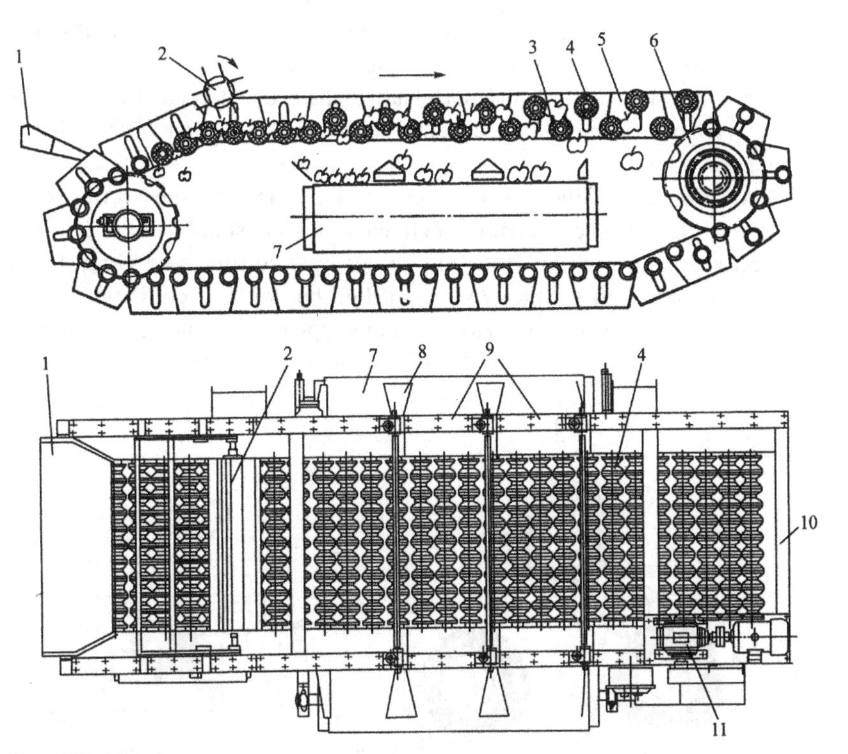

FIGURE 3.15 Three-roller fruit-sorting machine. 1. Material-feeding hopper; 2. roller; 3. sorting roller (fixing); 4. sorting roller (lifting); 5. connecting plate for the chain; 6. driving sprocket; 7. material-feeding conveyor; 8. partition board; 9. hoistway; 10. mechanical rack; 11. worm reducer.

bamboo-shaped roller, which is connected to both sides by chains. The rollers are divided into fixing rollers and lifting rollers. Fixing rollers are hinged with chains in the fixed position, while lifting rollers are mounted in a long hole in connecting plate 5 via a chain. The lifting rollers and fixing rollers on both sides form a series of diamond-shaped sorting holes (Figure 3.16). In addition, lifting slide 9 is designed for the roller on both sides of the chain.

During operation, the chain is driven by a gear to allow continuous running. Each roller undergoes continuous clockwise rotation because of the friction between the wheel and the slideways on both sides of the roller. Citrus fruits access the roller chain belt via the material-feeding hopper. The citrus fruits that measure less than the diamond-shaped holes will pass straight through the holes and fall into the collecting hopper. Larger citrus fruits are organized into a layer by the material-organizing drum. The citrus fruits then enter the grooves formed at the diamond-shaped hole location via the action of the lifting roller. The citrus fruits are transferred to the continuous sorting section sequentially. In this section, the hoistway has an inclined shape, which results in the gradual enlargement of the diamond-shaped holes due to the elevation of the roller. The citrus fruits keep rolling continuously into each hole while adjusting their positional relationship with the diamond-shaped holes because of the effect of friction with the roller shaft. When the citrus fruits aligned in a specific direction are smaller than the diamond-shaped holes, the citrus fruits drop onto the lateral transport conveyor and are transferred to the corresponding partition locations to pass along the conveyor. Larger citrus fruits move forward on the chain belt and are discharged at the end of chain belt if they still cannot pass through the diamond-shaped holes when the lifting roller is at its highest position.

This sorting machine has high productivity. During the sorting process, the citrus fruits can change their locations constantly relative to the diamond-shaped holes, which results in accurate sorting. The citrus fruits always maintain their contact with the roller, but without collisions so the fruits experience less damage. However, this type of machine has a complex

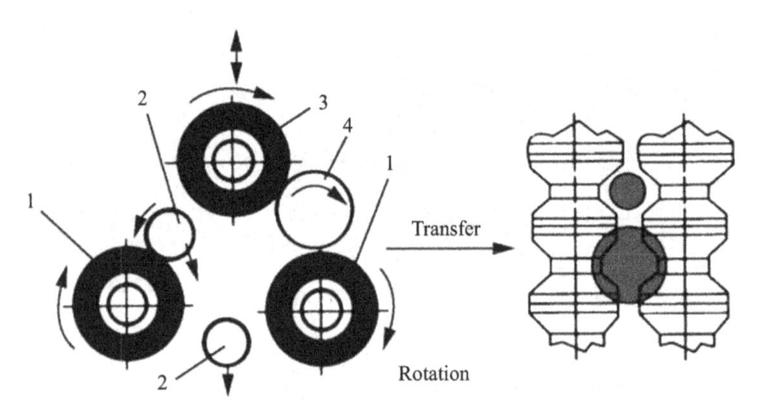

FIGURE 3.16 Schematic diagram showing the principle of three-roller sorting. 1. Fixed hierarchical roller; 2. raw material (small); 3. lifting roller; 4. raw material (large).

structure and high manufacturing cost. Thus, it is only suitable for large citrus-processing factories.

3. Citrus optical sorting machine

This machine uses a photoelectric sensor to detect the size of the raw materials. As the citrus fruits pass the photoelectric detector at a constant speed, the fruit height, diameter, and surface area can be calculated by detecting the beam-blocked time and beam-blocked number. After comparisons with the designed parameters, the discharging device can be controlled to allow citrus fruits to drop into the corresponding location to complete the effective sorting process.

(1) Beam-blocking citrus-sorting machine. Figure 3.17(a) shows a schematic of the two-cell shading-style sorting machine, where L is the light-emitting device and R is the receiver. A light emitter and a receiver compose a unit. The distance *d* between two units is determined by the sorting size required for citrus fruits. As the conveyor moves forward, the distance between the two units gradually becomes smaller. When the citrus fruits on the conveyor move forward and pass through the sorting zone, two light beams are blocked if the citrus fruit is larger than *d*. Next, the photoelectric element and control system induce the pulsing plate and nozzles to operate, thereby excluding specific citrus fruit via the lateral conveyor. The number of dual cells depends on the specifications required for citrus sorting. This sorting machine is only suitable for single-direction sorting.

(2) Pulse-counting citrus-sorting machine. Figure 3.17(b) shows a schematic of the pulse-counting sorting machine, where the light emitter (L) and receiver (R) are located at the upper and bottom positions, respectively, in the tray of the conveyor used for citrus fruits, and they are aligned with the middle

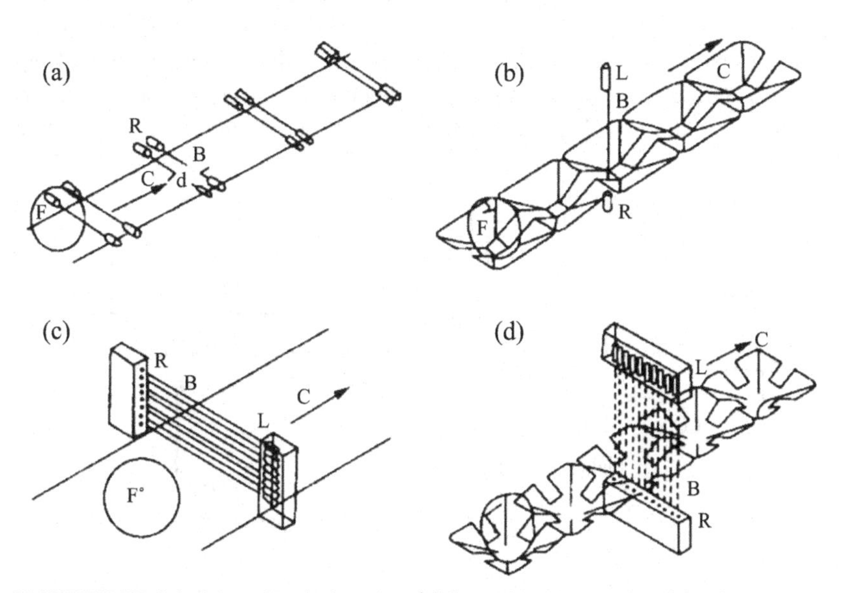

FIGURE 3.17 Principles of optical sorting. (a) Beam-blocking sorting; (b) pulse-counting sorting; (c) horizontal barrier sorting; (d) vertical barrier sorting.

of the tray opening. When the tray moves a distance a, the light-emitting device emits a pulse beam, and the number of blocked pulse beams is n during the movement of citrus fruits. The citrus fruit diameter is calculated as $D=na$. By computer processing, the value D is compared with the settings to determine the sorting specifications according to the size of the citrus fruits.

(3) Horizontal barrier sorting machine. As shown in Figure 3.17(c), the light-emitting devices and receivers are arranged in a row to form a light barrier. When citrus fruits move forward on the conveyor and pass through the beam barrier, the fruit height can be calculated according to the beam-blocked number. On the basis of the light beam-blocked time, the lateral projection area of the citrus fruit can be calculated by integration during the movement of citrus fruits on the conveyor. After comparison with the designed parameters, citrus fruits with different sizes can be excluded at different locations according to the corresponding specifications.

(4) Vertical barrier sorting machine. As shown in Figure 3.17(d), this machine is similar to the horizontal barrier sorting machine, but it detects the maximum size in the width direction and the cross-sectional area in the horizontal direction of the citrus fruit.

4. Citrus image-processing sorting machine

Mechanical dimension-based sorting is not strictly sorting based on the shape, but instead it is sorting according to size. Sorting based on the fruit shape requires the detection of fruit sizes in multiple directions, which allows the determination of the planar or three-dimensional shape of fruit. A computer-based image-processing sorting machine and its corresponding system configuration are shown in Figure 3.18. Compared with mechanical sorting, the most obvious difference is the application of a charge-coupled device (CCD) camera to allow noncontact imaging and shape judgment. Image-processing sorting machines have excellent stability and accuracy, as well as a powerful processing capacity during the application process, which can be used for sorting various citrus fruits. In practical applications, this type of sorting machine has a sorting

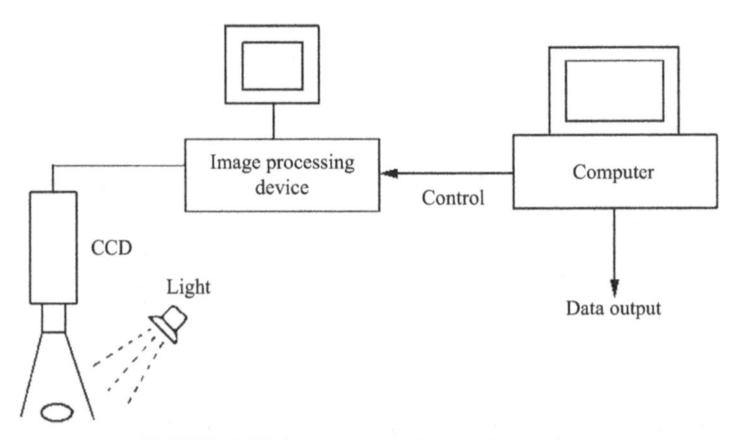

FIGURE 3.18 Image-processing sorting machine.

capacity of 3000 fruits per hour for large fruits and 10,000 fruits per hour for smaller fruits. The detection errors in terms of the diameter, length, or thickness of fruits can be less than ±1 mm. Highly detailed determinations can also be conducted. This type of sorting machine can determine the diameter, maximum diameter, and average diameter, as well as various coefficients related to abnormal fruit shapes.

The general procedures used for image processing are as follows. First, the image captured by the camera is converted into a two-dimensional image, and the length, width, and area are calculated, which are characteristic values used to sort fruits. Finally, the calculated characteristic values and designed reference values are compared to facilitate sorting. A more accurate method involves further thin-line and clear-shape feature processing or differential processing after calculating the characteristic values. This more accurate detection method can be achieved using graphic connectivity. Depending on the characteristic values used for sorting, if the characteristic value is far from the reference value, even if it is very small, the sorting grade can be changed, which results in a big difference compared with human judgment. To improve this approach, a fuzzy method has been developed for determining compatibility. If the characteristic values cannot be used to evaluate the fruit shape, an artificial neural network can be used to replace the characteristic values (diameter, length, and flatness) using a Fourier transform to evaluate the shape of the fruit.

Figure 3.19 shows the citrus quality-grading process. The grading operations comprise hierarchical arrangement, separation, camera analysis, damage detection, and the computer processing and judgment process. To ensure the accuracy of the image captured and to avoid the adhesion of citrus fruits,

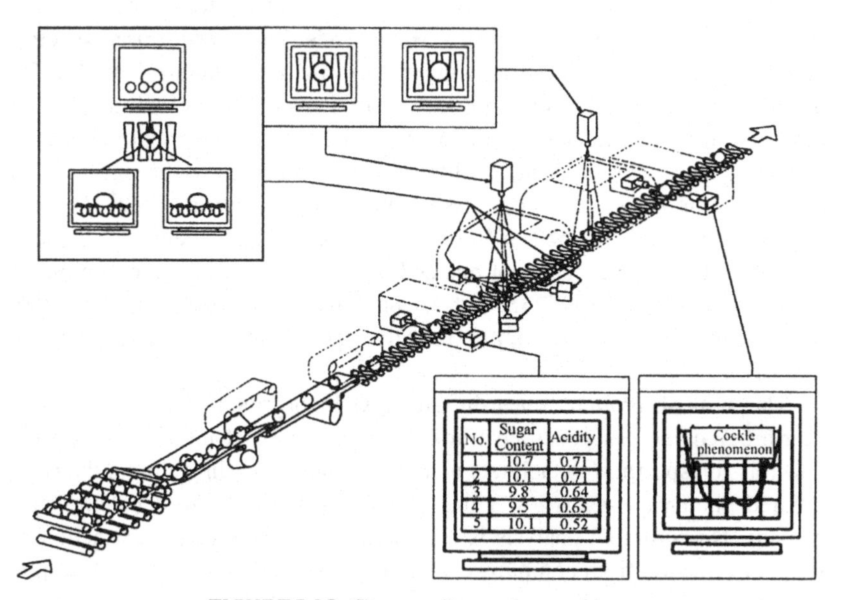

FIGURE 3.19 Citrus quality-sorting machine.

two conveyors operate at different speeds in a V-shaped arrangement to make the fruit form a single row, because the designed speed of the roller chain is higher than the conveyor speed, which facilitates the separation of the citrus fruits as they line up together. Five CCD cameras are arranged above and around the conveyor, which acquire images of the citrus fruits from all directions. However, a citrus fruit-flipping device is installed to acquire top and bottom images of the citrus fruits. In addition, a noninvasive detection device is installed on both sides of the conveyor to detect the sugar content, acidity, decay, injury, and wrinkled skin. As the citrus fruit pass the CCD camera, the color, size, shape, internal quality, sugar content, acidity, and surface damage status are all recorded. On the basis of the information acquired, the sorting process can be completed successfully after computer processing.

5. Inherent citrus quality-sorting equipment

1) Near-infrared sorting equipment

(1) Near-infrared analysis technology. An infrared ray with a wavelength of 0.8–2.5 μm is called a near-infrared ray. The near-infrared analysis method applies chemometrics to analyze ingredients and physical and chemical characteristics by near-infrared spectroscopy. At present, this is the most widely used method because it is a relative mature technique. Modern near-infrared spectroscopy has developed rapidly, and it is the most notable spectroscopic technique that has emerged since the 1990s. Near-infrared spectroscopy is based on nonharmonic oscillations from molecular vibrations, specifically the octave absorption and co-frequency absorption of the molecular vibrations from hydrogen-containing groups such as —OH, —SH, —CH, and —NH groups. Plant- and animal-derived foods contain large numbers of these groups; thus, the absorption spectra of these groups can be used to characterize the chemical structures of these components in various foods. The near-infrared spectral region can generate a large amount of information, which is suitable for measuring the sugar content and acidity of citrus fruits, as well as analyzing internal lesions in citrus fruits. For example, common components such as water, sugar content, and acidity may be reflected by the characteristic absorption of —CH groups. Experiments have confirmed that the Brix values determined by near-infrared spectroscopy and chemical methods exhibit excellent linear correlations and high consistency. Determinations at wavelengths of 914, 769, 745, and 786 nm have the highest accuracy with a correlation coefficient of approximately 0.989 and a standard error of 2.8 °Brix.

The weak absorption of near-infrared rays by organic materials means that near-infrared ray can penetrate fruits to acquire internal information via the transmission spectrum to facilitate noninvasive detection. In addition, the near-infrared photon energy is lower than that of visible light, so it cannot harm humans. Near-infrared analysis is a technique that requires the extraction of weak information from complex, overlapping, and variable spectra. Thus, it is necessary to establish a mathematical model based on modern chemometrics. It is essential to establish a model with high stability and precision before the application of near-infrared spectroscopy.

The establishment of near-infrared analytical methods involves four steps: selecting a representative calibrated sample and measuring its near-infrared spectrum; applying a standard or accepted method to analyze a specific component or characteristic data; establishing a calibrated model via rational chemometrics based on the spectra determined and basic data; and pretreating the spectra and determining the compositions of unknown samples based on the correlations between the spectra and basic data.

(2) Near-infrared sorting device for sugar acidity analysis. Near-infrared rays can be used to analyze citrus fruits, particularly to measure the sugar content and acidity. Figure 3.20 shows an online of a sorting device based on noninvasive determination of the sugar content or acidity in citrus fruits. This device consists of a light source, optical detector, and data-processing unit. Reflected light does not represent the internal status of citrus fruits, but the light transmitted by the internal part of the fruits can be acquired by an online detector to obtain its transmission spectrum. After photoelectric conversion and calculation, the sugar content and acidity are sent to each device. Citrus peel is thick; thus, it is necessary to increase the light power to ensure that the near-infrared light has sufficient energy for transmission through the citrus fruits. The sorting device can achieve a sorting speed of three to five citrus fruits per second, with a measurement error of <1 °Brix if the fruit diameter is 45–120 mm, the fruit

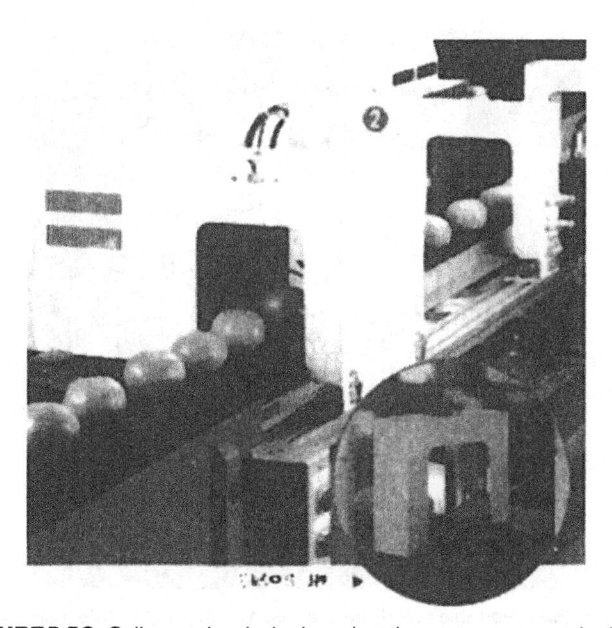

FIGURE 3.20 Online sorting device based on the sugar content and acidity.

height is >31 mm, and the minimum fruit interval is 10 mm. Under these conditions, this sorting device is suitable for detecting 16 different types of citrus. Using this device, the sugar content and acidity of citrus fruits can be determined rapidly without any damage. The sugar–acid ratio of citrus fruits can be displayed in addition to the sugar content and acidity. The pH can be measured, but there may be larger errors in the pH determination compared with the Brix value at low concentrations.

Figure 3.21 shows a portable sorting device for determining the sugar content and acidity of fruits. In contrast to the fixed device, which can be used for determining the sugar content, acidity, and quality grading, the portable sorting device based on the sugar content and acidity can also be used to monitor changes in the internal composition in real time and provide growth records during the growth of citrus fruits. The data obtained can be stored on a PC card for future analyses.

2) Ultraviolet (UV) sorting device

(1) UV analysis technology. UV radiation consists of wavelengths of 100–380 nm. At wavelengths of 320–380 nm, UV can convert chemical energy into kinetic energy to stimulate molecular movement. When fruits are damaged, the cells in fruit peel that contain essential oils are destroyed, and the essential oil is released onto the surface of the fruit. In a darkened room, the molecules can be excited from the ground state to the excited state by irradiation using UV light. By contrast, if the molecules recover to the ground state from the excited state, the irradiation energy will be released from the site of injury in the form of fluorescence. This fluorescence is

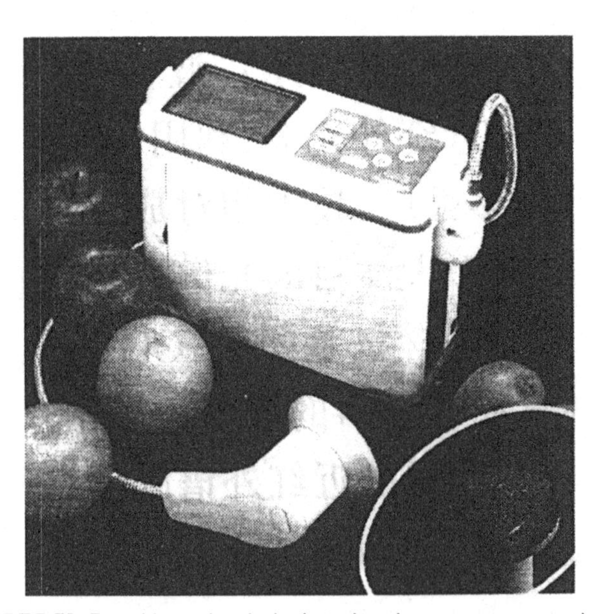

FIGURE 3.21 Portable sorting device based on the sugar content and acidity.

visible and this facilitates damaged fruit detection. By contrast, no visible light is observed from the normal sites in theory. Thus, there is an obvious difference in terms of light and dark between the normal and damaged sites. This is defined as the difference in reflection between normal fruits and damaged fruits under UV irradiation, where detection and sorting can be achieved on the basis of computer image processing.

(2) UV sorting device. Figure 3.22 shows a schematic diagram of an online UV sorting device for damaged citrus fruits. This device consists of a camera, UV lamp, conveyor, and lamp shade. The darkroom includes lamp shade and a rubber cover, which are used to block visible light and eliminate possible interfering factors. Fluorescent light can be used as the light source. To improve the light intensity, the lamp shade can be a semi-cylindrical shape that is painted white on the inner surface to increase the reflective effect. The image of the citrus fruit is captured in the center of the lamp shade. In addition, a small hole is made in the lamp shade to form the imaging path. The citrus fruit enters the darkroom and passes beneath the camera, where camera recording and computer image processing are used to facilitate sequential comparisons.

Figure 3.23 shows several images of the same citrus fruit during the UV sorting process. The image on the left was obtained under normal conditions. It is difficult to observe any damage, but obvious fluorescence is observed on the top of the citrus fruit under UV light irradiation (middle image), whereas no visible light is emitted by the normal area of the citrus fruit. The image on the right shows the computer-processed image where the white region is the damaged area of the citrus fruit. The white area is determined and the damage level of the citrus fruit can be assessed.

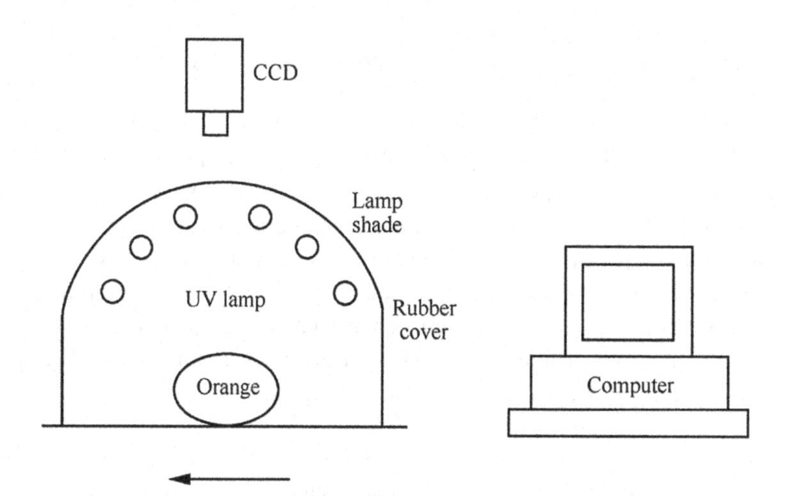

FIGURE 3.22 UV sorting device.

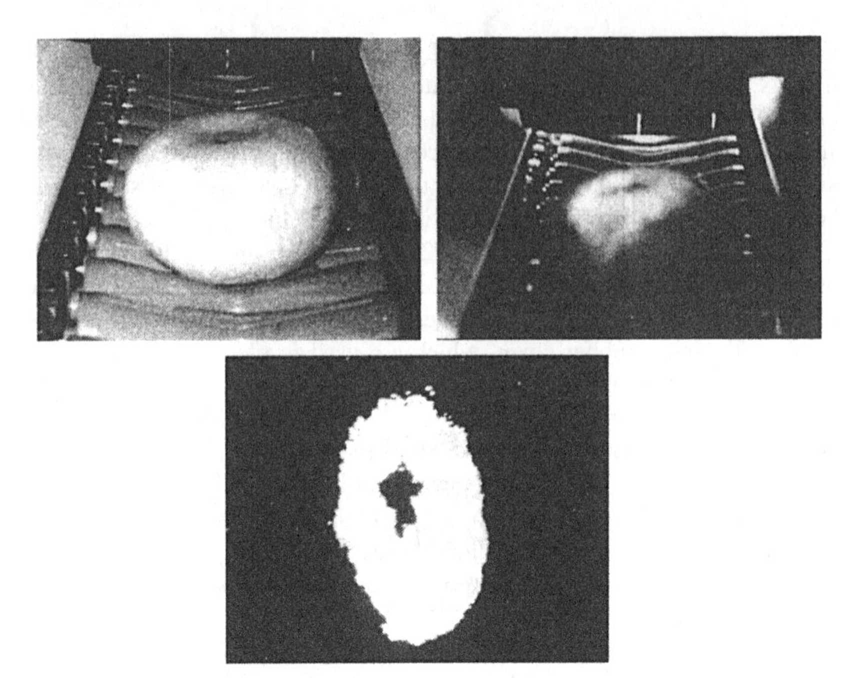

FIGURE 3.23 UV sorting to identify damaged citrus fruit.

The accuracy of detecting damage on the citrus fruit surface is affected by multiple factors such as the UV light source intensity, UV peak wavelength, distance between the light source and citrus fruit, citrus fruit damage status, citrus fruit temperature, and citrus species. The UV light peak wavelength can affect the fluorescence intensity. Therefore, a light source with specialized processing (less visible light) and a peak wavelength at 352 nm is the optimal light source. If the UV light source is too strong, however, the effect is not obvious. Thus, the optimal light source intensity should be 60 W. To reduce interference, the camera is usually equipped with a filter. The damage is determined by verifying whether fluorescent substances are present on the citrus peel. Some damaged citrus varieties may exhibit obvious fluorescence under UV light irradiation, whereas species such as lemon do not. It should be noted that some pesticides on the surface of citrus fruits can also emit fluorescence, where a white area is also observed after image processing. However, the luminous points, scattered distribution, and shape differ slightly from the fluorescence related to damaged sites on fruit, although it may be difficult to distinguish pesticide residues and damaged regions, which can lead to erroneous assessments.

3) X-ray sorting equipment

(1) X-ray analysis technique. X-rays have high penetrative capacities, but the depth of penetration is affected by the density of the material. Soft X-rays are X-rays at longer wavelengths, which have less energy and weaker penetration than standard X-rays. During the citrus fruit detection process, the requisite

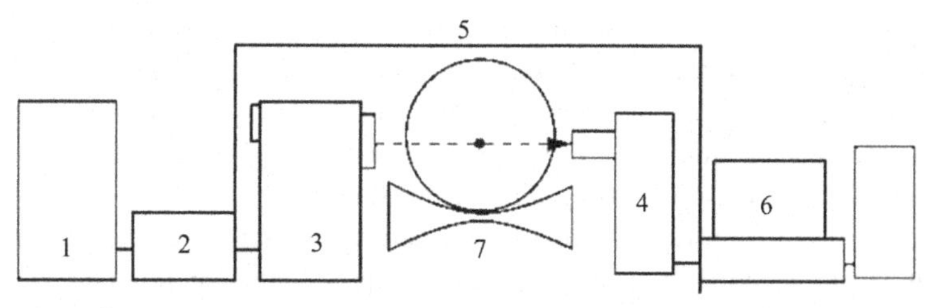

FIGURE 3.24 Schematic diagram showing the device for sorting citrus fruits with wrinkled skins. 1. Power; 2. control system; 3. x-ray-emitting device; 4. receiver; 5. hood; 6. computer; 7. conveyor.

X-ray intensity should be weak because the fruit density is low. Therefore, soft X-ray detection is suitable for practical applications.

Soft X-rays can be used to detect wrinkled skin and internal defects in citrus fruits. During the growth period of citrus, wrinkled skins (thick peel and a low pulp volume) often develop because of the impact of environmental conditions. Wrinkled fruits have a low water content and poor taste, which means that they are substandard. Therefore, they must be eliminated during grading.

(2) Sorting device for wrinkled skin citrus fruits. Figure 3.24 shows a schematic diagram of a device for sorting citrus fruits with wrinkled skin and the corresponding detection wave shape. The X-ray detection device consists of a transmitter, receiver, light-blocking cover plate, conveyor, and computer. To ensure the safety of the operator, a 2.3-mm-thick iron shield is necessary to exclude X-rays. The emission and receiving devices are arranged on both sides of the conveyor. When no citrus fruit is on the conveyor, all of the X-ray signals are received by the receiver and converted into electrical signals. After A/D conversion and inputting into the computer, the maximum signal value is observed at this time. As the citrus fruits pass through the X-rays, the signal value will fluctuate according to the density of the fruit.

The detection results obtained using this sorting device are not provided in the form of images, but instead a waveform shape is acquired to reduce the imaging time and to facilitate rapid detection. Normal fruits generate a waveform shape with a smooth transition, whereas wrinkled skin fruits produce a waveform shape with an abrupt change due to a decrease in the local density. Thus, the presence and absence of wrinkled skin citrus fruits can be detected on the basis of an abrupt change in the waveform.

During actual testing, accuracy is one of the critical determinants when selecting this technique. The detection accuracy is affected by many factors, such as detection position, citrus fruit size, X-ray intensity, conveyor speed, and the discriminant value. Wrinkled skin is a physiological phenomenon that occurs often near the stem and flower pedicle in citrus fruits. Thus, the X-ray scanning area and computational analysis should focus on both ends of the fruit, whereas the middle area can always be ignored. In addition, excessively strong or weak X-rays can cause difficulties recognizing abrupt changes in the waveform. Before testing, appropriate adjustments should be made depending on the actual situation. For example, the use of an excessively fast conveyor can reduce the detection accuracy, whereas a slow system can affect productivity. The choice of a reasonable discriminant value should be based on a large number of experiments.

According to the regulations of international organizations such as FAD/IAEA/World Health Organization, safety experiments are not necessary if the radiation dosage is <10 kGy, while all the requirements are satisfied if the absorption dosage applied by the system is ≤0.03 Gy.

3.3.2 Scraper-Style Continuous Citrus Peel–Heating Machine

The chain belt-type continuous citrus peel-heating machine is the main equipment used on a canned orange production line, where a chain-type conveyor is used to transfer raw citrus materials. A scraper is installed on the conveyor, so this type of machine is known as a scraper-style continuous citrus peel-heating machine, as shown in Figure 3.25, which mainly consists of water tank 7, steam-blowing pipe 4, chain-style conveyor, and transmission system.

Water tank 7 is the major component of the entire apparatus, which consists of 4-mm-thick welded stainless steel plate. The scraper 3 is welded onto the chain plate between two chains 8. To reduce the resistance of the scraper during

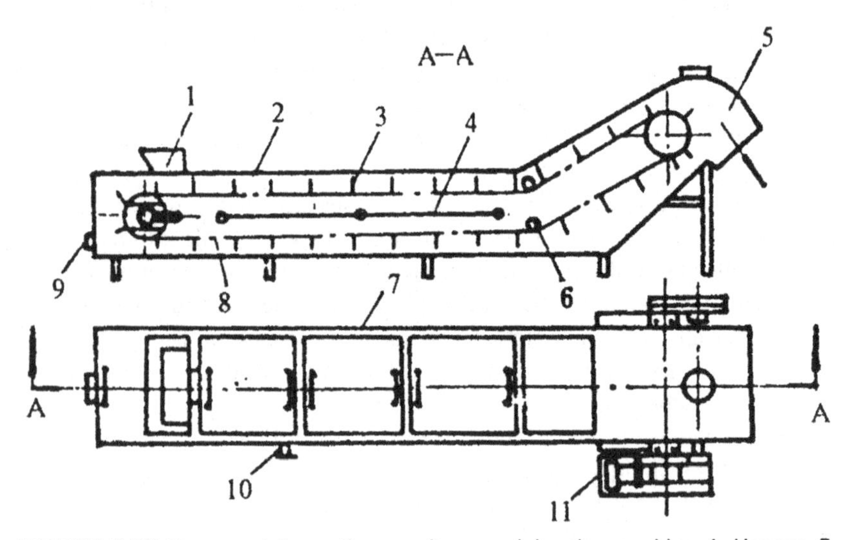

FIGURE 3.25 Scraper-style continuous citrus peel–heating machine. 1. Hopper; 2. head cover; 3. scraper; 4. blowing pipe; 5. material-discharging port; 6. pressing wheel; 7. water tank; 8. chain; 9. waste-discharging port; 10. overflow port; 11. speed-controlling motor.

operation, an appropriate number of holes are drilled in the scraper. The pressing wheel 6 can allow the chain to transit to an inclined segment to allow material discharge from the horizontal segment. The pressing wheel can also tighten the chain belt. Citrus raw materials are sent to the hopper 1 via a transporter and they drop onto the chain belt (or bucket bath), which is equipped with a scraper. The water tank is filled with water and heated by steam from steam-blowing pipes 4. The citrus peel is heated during the movement of the chain belt, which carries the citrus fruits from the material-feeding port to the material-discharging port. Bubble-blowing pipe 4 has numerous openings. There are more openings close to the material-feeding port and fewer near the material-discharging port, which ensures that the low-temperature raw materials receive preheating treatment at when they are fed into the machine. To reduce the direct impact of steam on the raw materials, there should be more openings on the sides and less on the top. This design accelerates the cycling of water in the tank and maintains a uniform temperature in the water tank compared with steam blowing with an uneven distribution of openings. Therefore, the steam pressure should be higher than 0.4 MPa, and the bottom of the water tank must be inclined to allow drainage. Overflow pipe 10 is designed to maintain a stable hot water level inside the water tank. The preheating duration can be controlled by adjusting the chain belt speed using the speed-controlling motor 11 or driving wheel. A lid 2 covers the top, which seals the edge of the seal groove inside the water tank to prevent steam leakage.

3.3.3 Orange Segment-Sorting Machine

The orange segment-sorting machine is the main equipment used for canned orange segment production, as shown in Figure 3.26. It is a continuous roller sorting machine that sorts orange segments into different sizes depending on the specifications.

This machine consists of a mechanical rack, material-feeding tray, grading roller, material-discharging slot, transmission system, and a CVT. The transmission system drives the roller via CVT-driven sprockets and gears.

Working principle: The orange segments are brought into the material-feeding tray by the conveyor. After rinsing with water, the cleaned orange segments are introduced into eight pairs of rollers and are sorted by tapered gaps between rollers, depending on the grading specification. The sorted orange segments are discharged from the unloading slot.

The gaps between the rollers are 3–11 mm at the narrow end and 19–27 mm at the wide end.

3.3.4 Pilot Equipment for Removing Citrus Sac Coating by Complex Enzyme

In the canned citrus processing industry, the acid–base methods were widely applied to remove the citrus sac coating from citrus segments in China. However, there are problems such as products broken and segments loosen

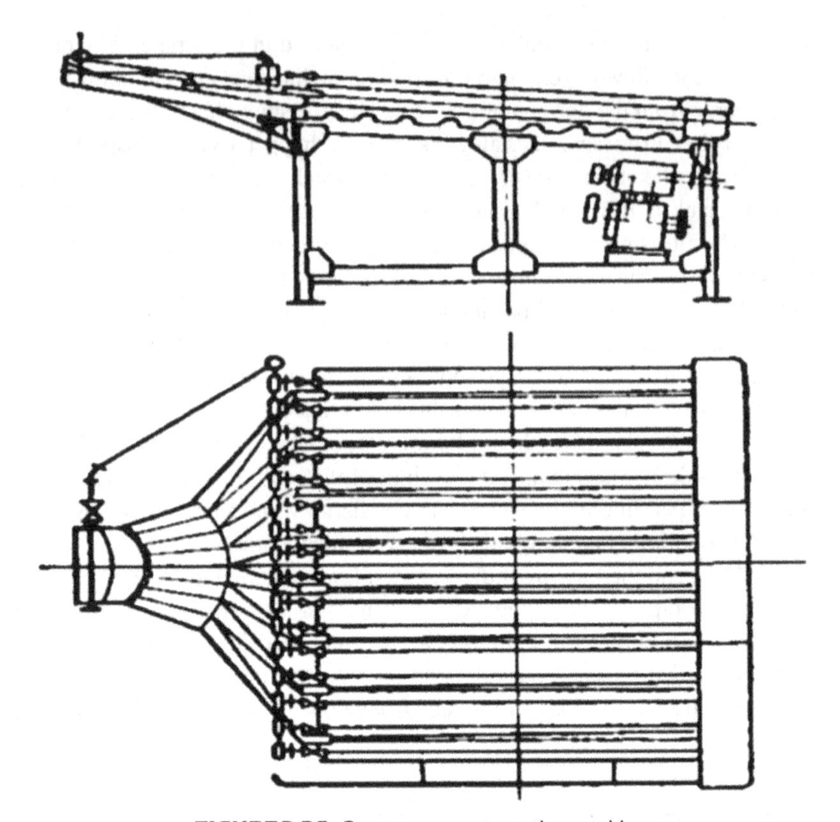

FIGURE 3.26 Orange segment–sorting machine.

easily, as well as the consumption of large amounts of water by traditional processing methods. Large amounts of acidic and alkaline wastewater are also generated by this process, which increase the production cost and degrade the product quality. The Hunan Agricultural Product Processing Institute developed a novel enzymatic process for removing the citrus sac coating as part of the "11th 5-year" plan. In this method, complex enzyme is used to hydrolyze the citrus sac coating of citrus segments. Compared with the acid or alkali methods used to remove the citrus sac coating, this novel method is characterized by high productivity, stable product quality, high safety, and no environmental pollution.

The structure of the pilot equipment used for the enzymatic removal of the citrus sac coating of citrus segments is shown in Figure 3.27, which consists of a mechanical rack, pumps, fine filters, control cabinet with a variable frequency-controlled chip, rushing water tank, coarse filter head, backwater tank, overflow pipe, backwater pipe, heaters, valves, and exhaust equipment. The rotation of the pump motor can be adjusted to 125–1250 r/min, the flux is 0.1–1.5 m^3/h, and the water temperature is 0–65 °C.

Operating principle: The water in the backwater tank is extracted by the water pump and transported to the bottom of the water-flowing tank, which has

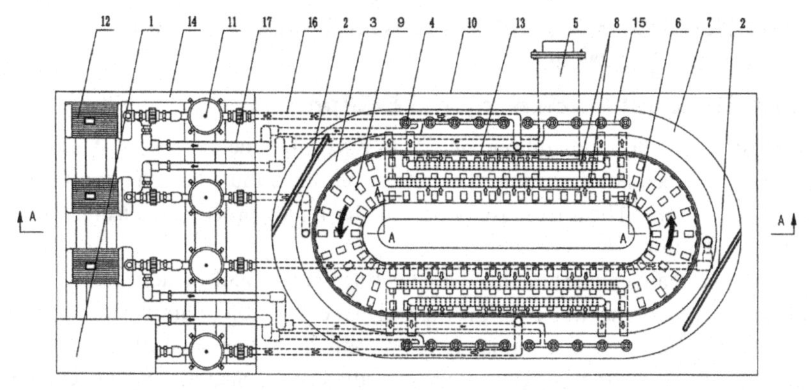

FIGURE 3.27 Pilot equipment for removing citrus sac coating by complex enzyme. 1. Control unit; 2. heater; 3. water storage chamber; 4. coarse filter; 5. material-discharging port; 6. circular working pool; 7. backwater tank; 8. overflow pipe; 9. circular working indicator; 10. shell; 11. fine filter; 12. water pump; 13. filter plate; 14. base; 15. connecting pipe for the coarse filter; 16. water inlet pipe; 17. backwater pipe.

an even distribution because of a 45° inclined outlet, after fine filtration. The filtered water flows from the outlet to form a circumfluence. In addition, a backwater equipment is installed in the bottom of the tank. The water is filtered by the coarse filter head and sent to the water pump via the backwater pipe to circulate. The overflow water in the water-flowing tank returns to the backwater tank. The flow rate of the water pump and the water temperature are controlled by the frequency-inverting controller and the temperature controller in the control cabinet, respectively. The remaining residues after enzymatic hydrolysis are discharged from the waste-discharging port.

The advantages of this equipment are as follows.

(1) It forms a circuit that ensures recycling, which improves the utilization rate for the treatment solutions.

(2) The circulation of the enzyme solution ensures that it makes full contact with the citrus segments, thereby accelerating the treatment speed, reducing the treatment time, and improving the treatment efficiency.

(3) The frequency-inverting controller changes the water pump speed to ensure the even flow of the circulating water and it reduces collisions between citrus segments.

(4) The system is equipped with an automatic temperature control equipment, which maintains the optimal treatment temperature for the treatment solution, thereby improving the citrus sac coating removal efficiency, reducing the soaking time, and improving the taste (and the brittleness) of fruits and vegetables, such as citrus fruits.

(5) The backwater tank has a coarse filter equipment that retains the peel and citrus sac coating from fruits and vegetables after treatment, which facilitates cleaning. Similarly, a fine filter equipment is installed in the water inlet pipe to improve the quality of the returning solution.

3.4 EXHAUST AND STERILIZATION MACHINERY AND EQUIPMENT

3.4.1 Exhaust Machinery and Equipment

Exhaust machinery and equipment aim to maximize the exclusion of air from canned products. After sealing, the canned products must have an appropriate level of vacuum. Thus, exhaust is one of the major processes during canned citrus production.

The two main exhaust methods are heating exhaust and vacuum sealing. The former uses steam heating to allow the thermal expansion of air in canned products and the evaporation of moisture from canned products to exclude air. After steam heating, sealing is performed immediately to ensure that the requisite level of vacuum is present in canned products after cooling. The latter process is also achieved using two methods. The first is the vacuum pump method, which applies a vacuum pump to release the air from a six-chambered vacuum sealing machine. It seals the canned product immediately in the vacuum environment, thereby ensuring that the requisite level of vacuum is present. The second method employs a steaming machine. During the application of this method, the level of vacuum is obtained by blowing steam into the canned products through a stream nozzle to exclude air. After the air has been excluded from the canned product, sealing is performed immediately to ensure that the appropriate level of vacuum is obtained.

The most widely used thermal exhaust machines that employ steam as the heating source include gear disc and chain-belt exhaust boxes.

1. Gear disc exhausting cabinet

Because of its large capacity, the gear disc exhaust cabinet is suitable for the production lines found in large canned citrus-processing factories. After the canned products are placed on the gear disc, they are transferred among subsequent gear discs via the driving force due to the rotation of the gear disc and rail. The canned products are heated during this long-distance movement to exclude air, before they are discharged from the material-discharging port for canned products after passing along a winding pathway.

The heated exhaust cabinet is usually double-ended with a precanned entrance end and an outlet end for sealing the canned products using a semiautomatic sealing machine.

The structure of the gear disc exhaust cabinet is shown in Figure 3.28, which consists of the cabinet body, bracket, teeth disc, guide rail, transmission device, and heating pipe.

1) Cabinet

The cabinet is rectangular in shape and rectangular openings are present at both ends to allow the discharge of canned products. To avoid condensation dripping from the cover of the cabinet, the cabinet has a sloping cover. To observe the working status in the cabinet at any time, the cover of the cabinet is divided into several small sections where any of the small covers

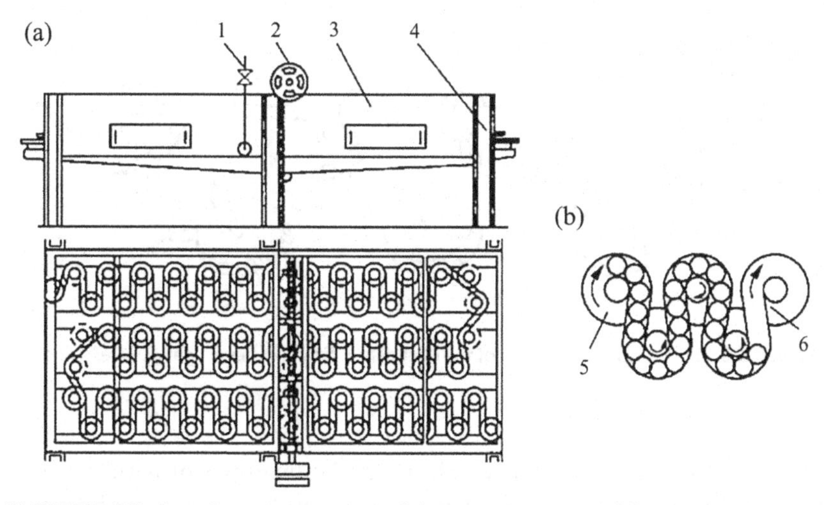

FIGURE 3.28 Gear disc exhaust cabinet. (a) Structural diagram; (b) movement pathway for canned products. 1. heating pipe; 2. transmission device; 3. cabinet body; 4. bracket; 5. gear disc; 6. rail.

can be opened as required. In addition, to exclude condensed water and to improve the water-sealing effect, the bottom edge of the cabinet and the four sides of the cabinet have grooves on their surfaces. During the heat exchange process, it is impossible to ensure that all of the vapor achieves condensation. Thus, to prevent uncondensed vapor diffusing into the work area from the rectangular openings at both ends, exhaust hoods are installed on both ends of the cabinet to release the vapor. A large amount of vapor is condensed into water during the heat exchange process, which is discharged from the bottom of the cabinet.

2) Gear disc and transmission system

There are 55–77 gear discs in the cabinet, which are divided into three groups with two rows in each group. A bracket is installed at each end of the cabinet to allow the discharge of the canned products. All of the gear discs must be installed on the same plane. Two adjacent gear discs are not engaged but the gear discs in the same group should be engaged. To ensure smooth operation and to increase the running distance for canned products, aluminum sheets are installed on top of the gear discs as a guide rail.

As shown in Figure 3.29, the transmission device is mounted on the bracket below the middle of the cabinet. The power of the motor passes through the pulley to drive the shaft rotation via a gearbox. The shaft is equipped with three bevel gears 2 to allow rotation in three axes, because the three bevel gears have different installation positions and directions. Two bevel gears on both sides have opposite installation directions, while the other bevel gear is in the middle. Therefore, the bevel gears run in a counterclockwise direction on both sides, whereas the bevel gears in the middle

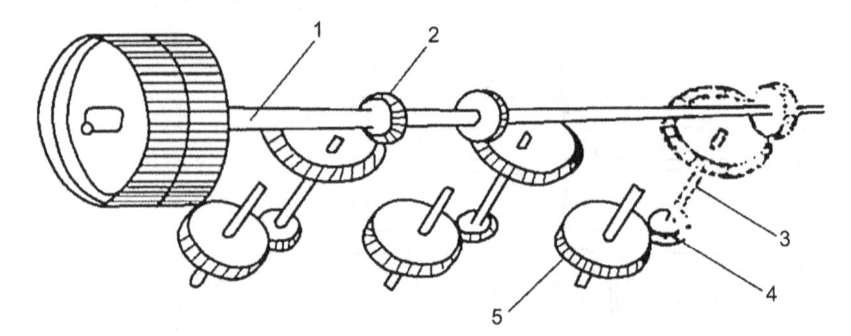

FIGURE 3.29 Transmission device of the gear disc. 1. Shaft; 2. bevel gear; 3. axle; 4, 5 gear.

run in a clockwise direction, which power the two rows of bevel gears in the middle and the four rows of bevel gears on both sides to allow their rotation.

3) Heating system

The heating system consists of the steam-inlet tube, dispensing tube, and three tubes in the longitudinal direction of the cabinet (a small hole is located at the top so steam can be injected directly into the cabinet).

2. Chain-belt exhaust cabinet

In this method, a chain provides traction for the conveyor machine used in exhaust devices. In the chain-belt conveyor, a cabinet is installed with a heating system. The governor is replaced with a CVT so it becomes a chain-belt exhaust cabinet.

The chain plate is welded directly on both sides of the chain so the chain can generate horizontal movements with the support of the rail. However, the rail must be installed at a horizontal level.

The ends of the cabinet are the entrance and discharge ports for canned products. There is no cabinet above the entrance and discharge ends of the chain. The length of the chain at the entrance for canned products is longer than that at the discharge end for canned products, which allows the access of canned products on the chain. The optimal height of the chain plate above the ground is 1 m, which may facilitate its operation.

Technical characteristics of the GT9B4 exhaust cabinet are:

Application scope	Can diameter: 70–120 mm. Can height: 40–130 mm
Exhaust time	6, 9, 12, and 15 min
Chain-belt speed	0.916, 0.611, 0.458, and 0.366 m/min
Motor J041-6	1 kW, 940 r/min
Dimension ($L \times W \times H$)	6550 × 980 × 1215 mm
Total weight	455 kg

3.4.2 Sterilization Equipment

Sterilization equipment is one of the most important components found in canned citrus-processing factories. The heat sterilization procedure conducted after canning products involves filling and sealing. At present, sterilization methods include pasteurization, high-temperature short-term (HTST) sterilization, and ultra-high temperature (UHT) short-term sterilization. Pasteurization is a low-temperature and long-term sterilization method where the sterilization temperature is below 100 °C (usually 62–65 °C) and the sterilization duration is 30 min. HTST sterilization usually requires a sterilization temperature below 100 °C and a sterilization duration of 15 s. UHT sterilization requires a sterilization temperature above 120 °C and a sterilization duration of only a few seconds. The HTST and UHT sterilization methods have high sterilization efficiency, but they also retain the excellent appearance, nutrition, and flavor of the sterilized products, where their performance is much better than that of other sterilization methods.

1. Horizontal sterilization equipment

Horizontal sterilization equipment is used only for high-pressure sterilization. This equipment has a higher capacity than vertical sterilization equipment. Therefore, this equipment is suitable for the production of canned oranges in large or medium-sized canned product factories.

A horizontal sterilization device is shown in Figure 3.30. The closure mode of the sterilization chamber 17 and the door 14 is similar to that found in vertical sterilization equipment. At the bottom of the sterilization chamber, there are two parallel tracks that allow the entrance and discharge of a cart that carries the canned products. The steam used to heat the products enters from the bottom of sterilization chamber via small holes in two parallel steam distribution pipes.

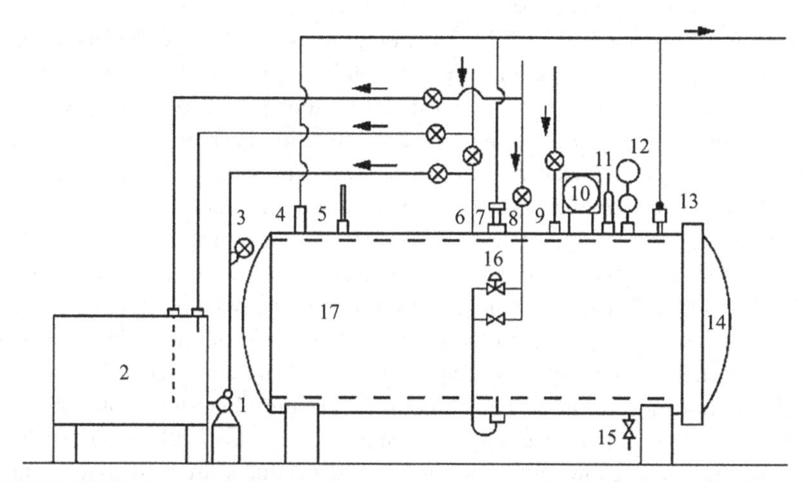

FIGURE 3.30 Schematic diagram showing a horizontal sterilization device. 1. Water pump; 2. water tank; 3. overflow pipe; 4, 7, 13. vent pipes; 5. safety valve; 6. inlet pipe for water; 8. inlet tube for vapor; 9. inlet pipe for compressed air; 10. temperature recorder; 11. thermometer; 12. pressure gauge; 14. sterilization device door; 15. drainage pipe; 16. film valve; 17. sterilization chamber.

The steam pipes are hidden under the rail. When the rail is parallel to the ground, the launch vehicle can be pushed into the sterilization chamber. Therefore, part of the sterilization chamber should be below the ground. A trough is required to allow drainage of the sterilization equipment.

The sterilization device is equipped with various instruments and valves. The application of back-pressure sterilization means that the pressure value in the pressure gauge indicates the pressure of the steam in the sterilization chamber and the pressure of compressed air; thus, the temperature and pressure are not in agreement. Therefore, a thermometer needs to be installed.

The afore-mentioned sterilization devices use steam as the heating medium. During operation, the presence of air in the sterilization chamber can result in an uneven temperature distribution, which affects the sterilization process and the quality of the product. To avoid "cold spots" due to the presence of air, the exhaust method is used during the sterilization process. The sterilization efficiency is improved by excluding air from the sterilization chamber by injecting steam through the exhaust valve on top of the sterilization device, which improves the steam flow.

2. Rotary sterilization device

This device is an HTST sterilization device, which reduces the sterilization time by improving the transfer rate in the heating medium. The rotational state is maintained during the sterilization process. The overall sterilization process is controlled by an automatic control system. The major parameters of the sterilization process, such as the pressure, temperature, and rotational speed, are recorded and adjusted automatically. However, this sterilization device operates as a batch system because it cannot allow the continuous discharge of canned products during the sterilization process.

(1) Structure. A rotary sterilization device with an overheated water circulation is shown in Figure 3.31. The device consists of two cylindrical pots. The upper pot is a water storage pot, which is a cylindrical sealed container that is used to prepare hot water for the bottom pot. The bottom pot is a sterilization pot with a twist. When the sterilization basket is placed into the sterilization pot, the installed pressure device restricts the relative movement between the sterilization basket and twist. The upper pot and the bottom pot are equipped with a liquid level controller. A transmission system is also installed in the sterilization pot, which consists of a motor, separable cone wheel-style CVT, and gear. The gear shaft drives the twist rotation on a fixed axis. The twist rotation results in the simultaneous rotation of the sterilization basket. The speed can be varied continuously within a range of 5–45 r/min and rotation may occur in two directions. During alternative rotation, the turning, stopping, and reversing movements are controlled by a time relay. During the sterilization process, the canned products are first loaded into the sterilization basket and separated by a soft pad between each layer. The sterilization basket is sent to the sterilization pot via a sterilization car on the track and fixed using a pressing unit. Finally, the motor drives the rotation of the sterilization basket.

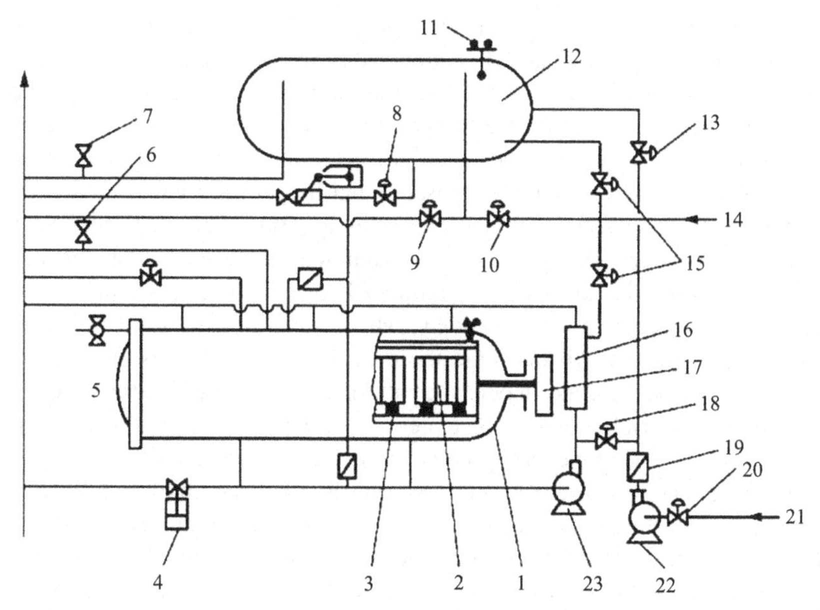

FIGURE 3.31 Schematic diagram showing a rotary sterilization device. 1. Sterilization pot; 2. sterilization basket; 3. rail; 4. water-releasing valve; 5. overflow valve; 6, 7. safety valves; 8. connecting valve between upper pot and bottom pot; 9. pressure-reducing valve; 10. pressure-boosting valve; 11. liquid level controller; 12. water storage pot; 13. hand valve; 14. steam; 15. heating valve; 16. steam and water mixer; 17. transmission system; 18. cold water throttle; 19. one-way valve; 20. cold water valve; 21. cold water; 22. cold water pump; 23. circulating pump.

The upper pot and the bottom pot are connected by pipes equipped with valves. A steam pipe, water inlet pipe, drainage pipe, and pneumatic tube connect the two pots at appropriate locations. Pneumatic, manual, or electric valves with different specifications are installed on these pipes, depending on the specific application. A circulating pump is installed on the sterilization pot to improve the heat transfer efficiency and to produce a uniform temperature via water circulation. In addition, the sterilization pot is equipped with a cold water pump that injects cold water into the water storage pot and cooling water into the sterilization pot.

(2) Working process. The working procedures for this device during one sterilization cycle are divided into eight operation programs. Each operation program is equipped with a series of number or light indicators. The operation programs are described as follows.

i. Preparation of overheat water. When the water level in the reservoir reaches a certain position, the liquid level controller stops the cold water pump. At the same time, the heating valve is opened in the water storage pot, and the water in the storage pot is heated quickly by injecting 0.6 MPa steam. During heating, the rate at which the temperature increases is controlled at 4–6 °C/min. When the water reaches the designated temperature, heating is stopped automatically. The overheated water is retained for injection into the sterilization pot.

ii. Water supply to the sterilization pot. The sterilization basket is loaded into an autoclave and the autoclave is sealed. The automatic operation program is initiated by pressing the start button once. Before entering the second program, the valve that connects the pots is opened automatically. The overheated water in the water storage pot then enters the sterilization pot because of pressure and height differences. When the water level in the sterilization pot reaches a certain height, the connecting valve between the pots is closed automatically. After the connecting valve has been closed, it is reopened to suit different requirements and there is usually a 1–5 min delay, which maintains the pressure inside the sterilization pot. At this time, the pressure in both pots is closed. The length of the delay time depends on the canned packaging style and the packaging materials. Packaging materials with poor thermal conductivity will result in slow temperature increases in canned products, thereby requiring a long time to achieve pressure equilibrium between both pots. Therefore, the delay should be longer. Otherwise, the bottle cap might be flattened by the pressure, which is also an issue with aluminum containers and soft canned products. By contrast, tinplate cans require a shorter delay time.

iii. Heating. When the overheat water come into contact with canned products in the sterilization pot, the temperature of the water begins to decrease and that of the canned products begins to increase via heat exchange. To reach the designated sterilization temperature, the heating valve on the sterilization pot is opened to inject steam into the sterilization pot through the steam–water mixer, which facilitates a rapid temperature increase in the sterilization pot. During the heating process, the twist begins to rotate and the circulating pump is opened to improve heat transfer efficiency by enhancing the water circulation.

iv. Sterilization. The sterilization process requires the maintenance of the designated temperature for a specific duration. Therefore, the heating valve on the sterilization pot must be opened to maintain steam injection and the continuous running of the circulating pump. After sterilization period is complete, the cooling program is started automatically.

v. Hot water recycling. After sterilization process is complete, the cold water pump operation commences. The cooling water is injected into the sterilization pot through a valve and the mixer is started. At the same time, the hot water in the sterilization pot is pushed back into the water storage pot. When the water storage pot is full, the connecting valve is closed immediately. Meanwhile, the heating valve of the water storage pot is opened to prepare the overheated water once again.

vi. Cooling. The cooling process usually employs an alternative cooling method, which includes enhanced-pressure cooling and reduced-pressure cooling, as well as reduced-pressure cooling alone, depending on the product requirements. The cooling method is chosen via a control panel based on preset options.

Enhanced-pressure cooling allows the canned products in the sterilization pot to be maintained at a stable original pressure, which is also called the reverse-pressure cooling method. During the enhanced-pressure cooling process, the cooling water pump operates and the throttle valve is in the throttled state so the pressure is maintained in the sterilization pot by controlling the cooling water with the throttle valve. When the enhanced-pressure cooling period is complete, reduced-pressure cooling is initiated. At this point, the cold water pump and circulating pump continue to run, but the drainage valve and overflow valve are opened. The butterfly valves that connect to the cool water drainage valve under reduced pressure are opened gradually because of the pulse, which generates a regular decrease in the pressure in the sterilization pot. After the complete sterilization device program is finished, the pressure in the sterilization pot is restored to the normal pressure and the overall cooling process is over. If enhanced-pressure cooling is not necessary, the clock for the enhanced-pressure cooling process can be set to zero.

vii. Drainage. After cooling, the circulating pump and the cooling water pump are stopped, and the water inlet valve is closed. Next, the overflow valve on the sterilization pot is opened to allow exhaust, and the cooling water is discharged rapidly from the cooling water drainage valve.

viii. Pot opening. After the discharge of the cooling water from the sterilization pot is complete, the sterilization device emits a signal so the sterilization basket can be removed.

3. Rotary continuous pressure sterilization equipment

To meet the demands of automatic continuous production, a rotary sealing valve is applied to connect the preheating pot, pressure sterilization pot, and cooling pot, which are combined to produce rotary continuous pressure sterilization equipment, as shown in Figure 3.32. This combination of equipment

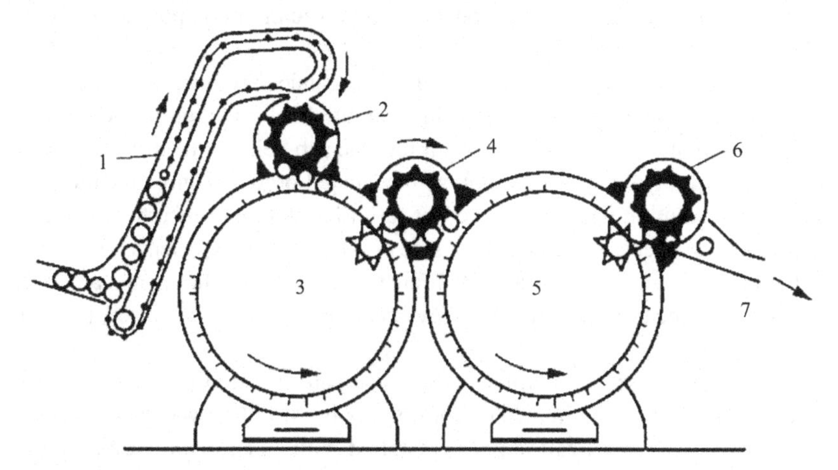

FIGURE 3.32 Sectional view of a rotary continuous pressure sterilization device. 1. Hoist; 2. inlet slewing ring of the pot; 3. heating sterilization pot; 4. transit rotary sealing valve; 5. cooling pot; 6. outlet slewing ring of the pot; 7. turret.

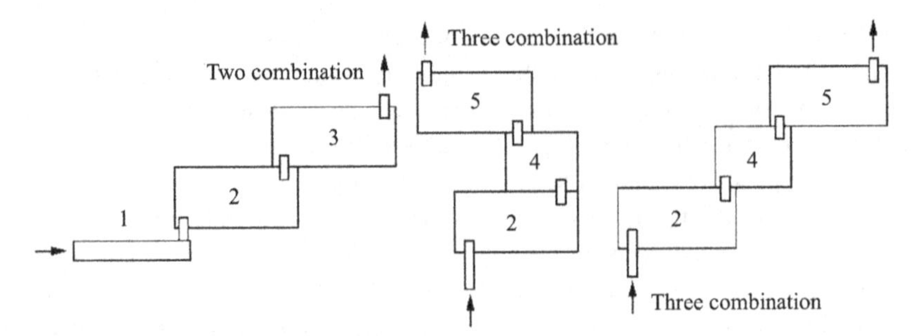

FIGURE 3.33 Pot combination styles. 1. Can-transporting machine; 2. pressure sterilization pot; 3. atmospheric or enhanced-pressure cooling pot; 4. reduced-pressure cooling pot; 5. atmospheric pressure cooling pot.

should have at least one sterilization pot and a cooling pot. The combination used depends on the product type and the sterilization process conditions. Several typical combinations are shown in Figure 3.33.

The preheating pot, pressure sterilization pot, and cooling pot have cylindrical bodies with pressure-resistant capacities of 0.2–0.3 MPa, where the diameter is 1473 mm and length is 3340–11,250 mm. The specific length and size depend on the sterilization duration, production rate, and container volume. Each pot has a rotary sub-grid canning rack and a T-screw rail is fixed along the pot wall. The canned products are fed from one end of the sealing pot through the rotary sealing valve and sent to sub-grid in the turret. The canned products are rotated together with the turret, where they undergo self-rolling and sliding on the spiral rail until the canned products reach the other end of the sterilization pot. The canned products are transferred sequentially through a rotary sealing valve to the next pot where they undergo continuous-heating treatment. Finally, the canned products are removed after completing the overall sterilization process.

Figure 3.34 shows the self-movement of the canned products in the sterilization pot. The canned products move in a longitudinal direction in the pot and reach the top of the pot, where they fall into the sub-cells of the turret. There is no contact with the walls of the pot, so the canned products only undergo synchronous revolution with the rotation of turret but without self-rotation. When the canned products are located at the side of the pot, they experience revolution and a small amount of self-rotation (sliding rotation). When the canned products are located at the bottom of the pot, they experience revolution and self-rotation (rolling).

The preheating pot and enhanced-pressure sterilization pot generally use steam as the heating medium. The cooling pot uses cooling water as the cooling medium. The sterilization pot is installed with an automatic temperature control device and the cooling pot is equipped with a liquid level control system.

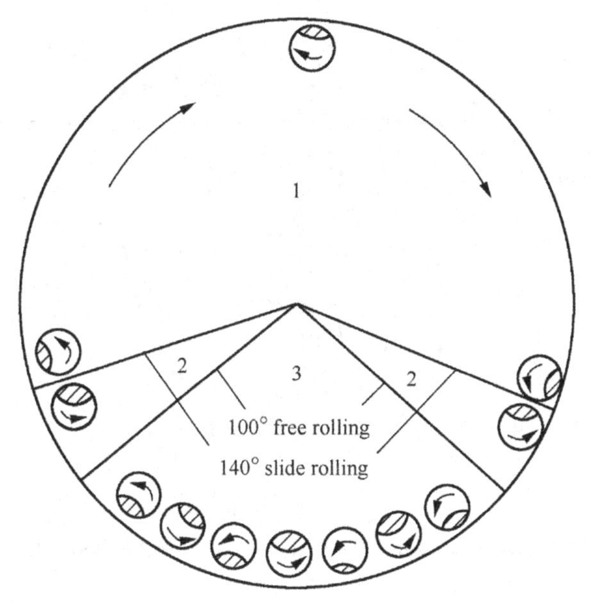

FIGURE 3.34 Diagram showing the movement of canned products in a rotary continuous sterilization pot. 1. Can-carrying area of the turret; 2. sliding-type rolling zone; 3. free rolling zone.

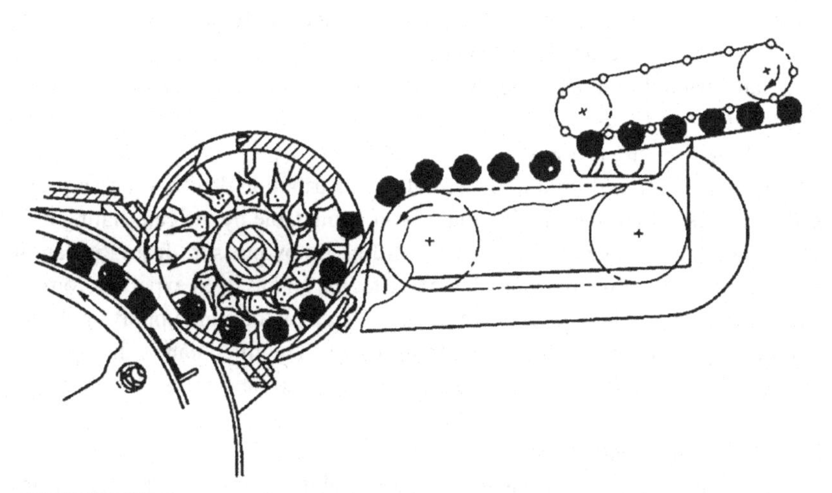

FIGURE 3.35 Rotary sealing valve located at the can-entry port in the sterilization pot.

The rotary sealing valve is the critical component of continuous rotary sterilization equipment. It allows the entrance and discharge of the canned products from the sterilization equipment and it also has the sealing function that maintains the operating pressure in the sterilization pot, as shown in Figure 3.35. Figure 3.36 shows how the transit rotary sealing valve facilitates the movement of the canned products from one pot to another pot.

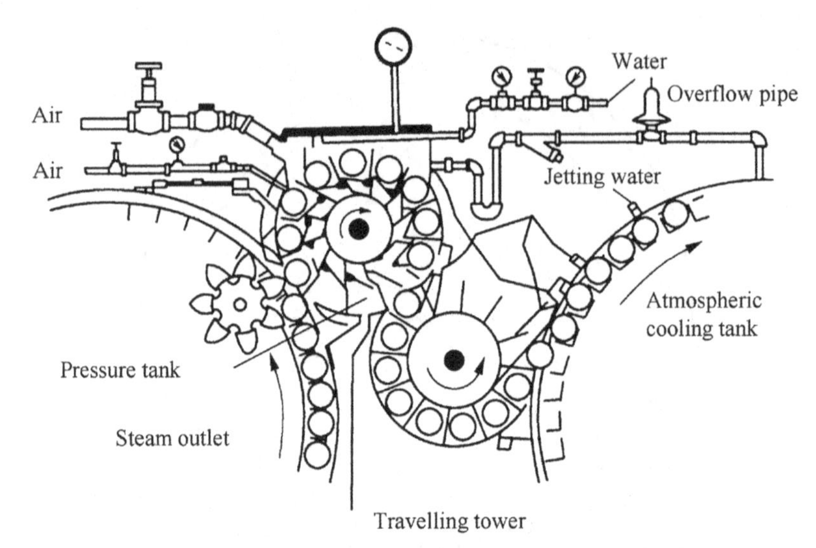

FIGURE 3.36 Transit rotary sealing valve located in the sterilization pot.

3.5 PACKAGING MACHINERY AND EQUIPMENT

3.5.1 GT786 Automatic Vacuum Juice-Filling Machine

The GT786 automatic vacuum juice-filling machine has 16 cup-style juice-filling tips. It is mainly used for adding sugar solution, fresh water, and salt solution to fruit and vegetable canned products. It is also used for fruit juice material loading, as shown in Figure 3.37.

1. Material-loading process

   The cans are transferred to the spiral can distribution device via conveyor 1 and then passed onto the rotating material-loading table by the can-dialing disc. The cans are moved upward to the juice-filling tip by pressure from the cam in the bottom of the can-supporting device. After the air has been expelled from the can and the can has been filled with juice, the cam on the can-supporting device moves downward and leaves the juice-filling tip. The filled can is then pushed out through the can-dialing disc and transferred to the can-sealing system by the conveyor.

2. Major components

   The juice-filling part consists of sugar solution storage, 16 juice-filling tips, and the liquid level control system. The juice-filling tips and volume cups are shown in Figure 3.37. When the core of valve 2 is at the start location, the juice-filling tips and volume cups are not connected with the vacuum system and the sugar solution system. When the body of can 5 presses the core of the valve, cap 4 rises and hole 6 points toward air-releasing port 7. At this point, the air is exhausted and the core of valve rises continuously. Volume cup 1 rises to leave the liquid surface. After hole 9 leaves air-releasing port 7, it is disconnected from the vacuum system. Meanwhile, hole 9 and hole 8 are connected so the sugar solution is injected into the cans through volume cup 1. Next, the material-filling tips return back to their original locations via the pressure of the

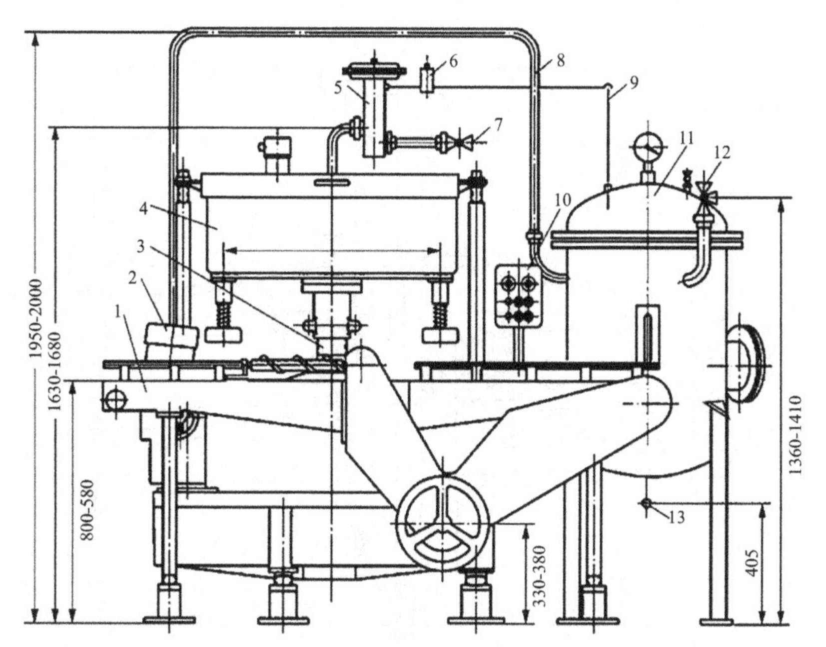

FIGURE 3.37 Schematic diagram of an automatic vacuum juice-filling machine. 1. Conveyor belt for the entry and exit of canned products; 2. transmission part; 3. central disc; 4. juice-filling part; 5. film valve; 6. solenoid valve; 7. inlet pipe for sugar water; 8. stainless tube; 9. copper pipe; 10. electrical panels; 11. steam storage tank; 12. pumping tube; 13. sugar solution outlet.

spring. Volume cup 1 is located above valve core 2. If the can does not contain any material, the volume cup is immersed in the sugar solution storage tank. When the volume cup is full, it is lifted so it leaves the liquid surface. According to the production requirements, the height of the volume cup can be adjusted to change the loading amount, thereby meeting the requirements for different volumes in the cans.

3. Central part of the disc

The central part of the disc consists of the can-supporting cage, vacuum distribution system, cam, and shaft, as shown in Figure 3.38. Spindle 10 is driven by gear 11 to rotate the big wheel, which rotates 16 can-supporting cages 5 on the slide of cam 8, as well as allowing rise and fall in specific positions. The cans can make the juice-filling tips rise and fall under the driving force of the can-supporting cages to allow evacuation and sugar solution filling.

3.5.2 Sealing Machinery for Canned Citrus Segment Products

1. Presealing machine

The presealing machine plays a critical role in the preliminary installation of the lids on the cans filled with materials. Appropriate tightness must be achieved, which means that the can cannot be opened by hand but the tightness should be suitable for the exhaust of air from the cans. The presealing machine

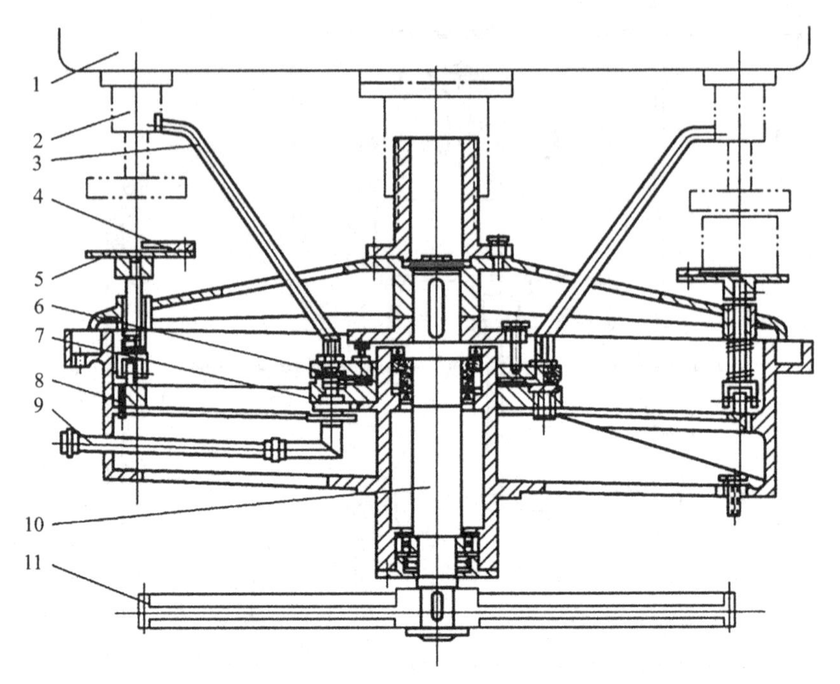

FIGURE 3.38 Central plate of the juice-filling machine. 1. Reservoir barrel; 2. juice-filling tip; 3. food use hose; 4. can-repelling board; 5. can-supporting cage; 6. vacuum allocation system (moving plate); 7. vacuum distribution system (fixing plate); 8. cam; 9. evacuation tube; 10. spindle; 11. Transmission gear.

is usually combined with a thermal air-exhausting machine or a high-speed vacuum-sealing device.

The overall device, as shown in Figure 3.39, consists of presealing heads, a lid-transferring machine, labeling system for the cans, electrical system, vacuum system, and conveyor system.

The filled cans are brought to the can-distributing spiral disc 6 at a specific distance interval by the can-transferring belt 7. The can is transferred to the lower tray 15 by the can-transferring dial 5 and the star-like dial 14 at the central kingpin 14. The can makes contact with a pressing board after the rotation of the can-distributing spiral. The pressing plate and connecting rod induce the lid-transferring system 8, as well as the connecting rod 18 and lip-sending rod 19, to push the lid out of the lid bore 20. The lid is then typed by the labeling system 9, which matches star-like dial 14 and the can on the lower tray. The top pressing head 11 starts to decrease with the support of the cam and the top pressing head 11 is connected to the vacuum system. The lid is maintained by the vacuum system and it remains in a specific position. The top pressing head 11 rotates to cover the lid on the can. Under continuous rotation, the can is moved to the curling arc plate 10 to complete the curling between the can and the lid. The settlement of the curling arc plate allows the diameter of the lid to be reduced by 2 mm at the exit of curling. The can spins around its own central line. When

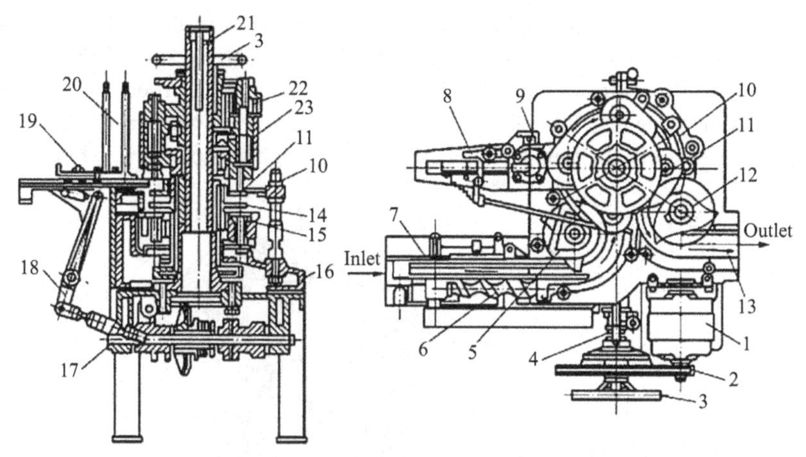

FIGURE 3.39 Presealing machine. 1. Motor; 2. belt and pulley; 3. hand wheel; 4. clutch; 5. tank-transferring dial; 6. can-distributing spiral disc; 7. can-transferring belt; 8. lid-transferring system; 9. labeling system; 10. curling arc plate; 11. pressing head; 12. can-releasing dial; 13. can-releasing belt; 14. star-like dial; 15. lower tray; 16. lower base; 17. horizontal shaft; 18. connecting rod; 19. lip-sending rod; 20. lid bore; 21. kingpin; 22. star-like gear; 23. top base.

the can leaves the curling arc plate, the pressure head rises and the filled can is transferred to the can-releasing belt 13 with the help of the can-distributing dial 12, before moving to the next machine.

2. GT482 automatic vacuum-sealing machine

This machine is a type of automatic vacuum-sealing machine with a single head and two pairs of crimping rollers. At present, it is used widely in canning factories in China for the vacuum sealing of various round metal cans.

This machine consists of an automatic can-transferring system, automatic can-distributing system, automatic lid-curling system, crimping head, can-unloading unit, transmission system, vacuum system, and electric control system, as shown in Figure 3.40.

After the cans have been transferred to the machine, they are spaced at a specific distance interval by the can-distributing spiral rod and passed to the transposition with a six-can spacing using an intermittent rotation style. The lid-distributing system can also transfer the can body to the slot with the lid. During the rotation of transposition, the can body will move the can-sealing position to remove the air from the can. After being sealed, the can is carried from the machine by the transfer belt. During crimping, the body of the can is not moved and the rotary wheel on the can mouth performs eccentric sealing in the crimping mode.

The vacuum degree in the sealed chamber is maintained by the vacuum system, as shown in Figure 3.41. Vacuum system mainly consists of the water-cycling vacuum pump 5 outside the sealing area, vacuum stabilizer 3, pipes 6, and sealed chamber 1 on the head of the machine. The vacuum stabilizer plays an important role in maintaining the stability of the vacuum during the sealing process. It may also separate the extracted liquid to avoid contamination of the vacuum pump during the evacuation process.

This machine also has various safety control devices, which ensure that the lid does not fall as the transferring belt passes through without cans, while cans

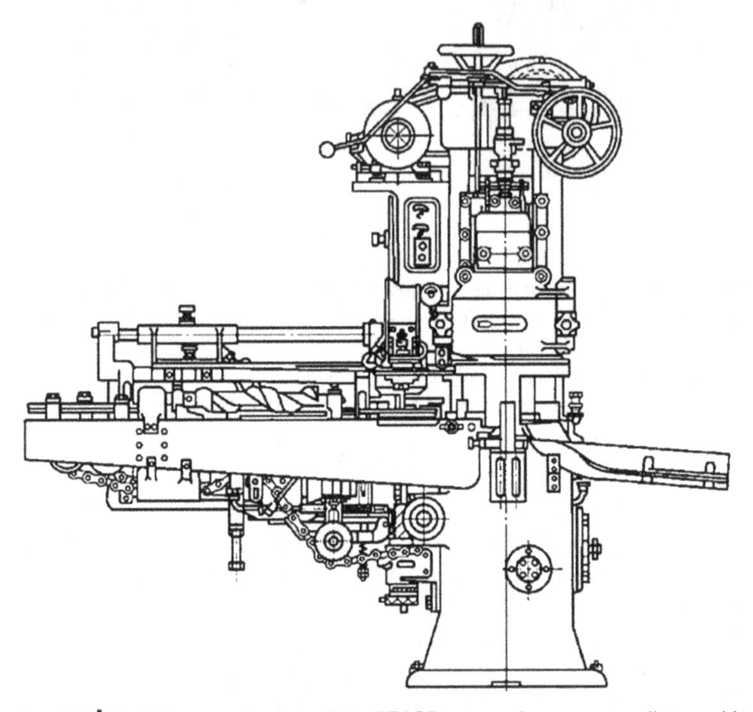

FIGURE 3.40 Schematic diagram of the GT482 automatic vacuum-sealing machine.

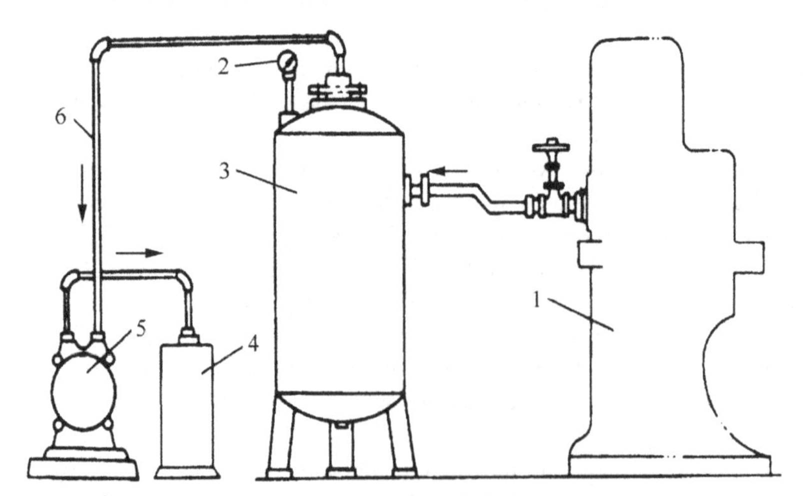

FIGURE 3.41 Schematic diagram of the vacuum system in the GT482 vacuum-sealing machine. 1. Sealed chamber on the head of the machine; 2. vacuum gauge; 3. vacuum stabilizer; 4. separator; 5. vacuum pump; 6. pipeline.

without lids can pass through the curling channel without any damage. In addition, if the machine has problems, it is stopped automatically. It is also suitable for sealing of round metal cans with various specifications by changing the type of curling head.

3.5.3 Spinning Capper

Spinning capping is a form of sealing for containers with threads or bayonets, where caps with threads or protruding teeth are screwed by special machinery. During canned orange processing, this type of machine is used widely to seal glass containers and plastic bottles. This type of seal is more convenient for unsealing and the caps can be recycled.

Spinning sealing is simple using a spinning capper, which consists of a bottle-supplying system, cap-supplying system, mechanical cap-screwing head, and position-controlling system. Figure 3.42 shows a claw-type cap-screwing head, which uses three claws to clamp a cap. Under the rotation of

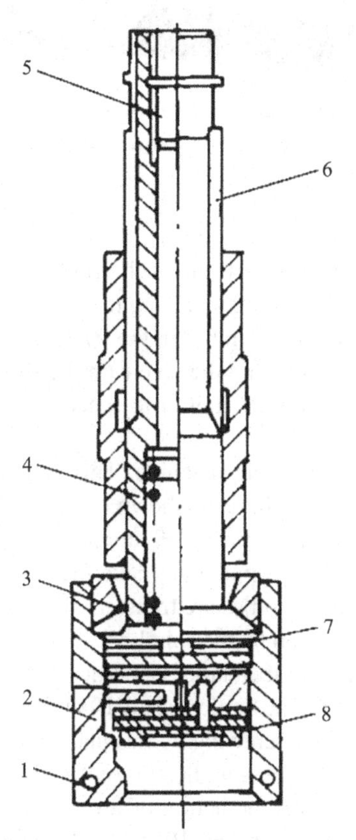

FIGURE 3.42 Structure of the capping head. 1. Screwing claw spring; 2. claw of screwing cap; 3. ball joint; 4. pressure-adjusting spring; 5. adjustable screw; 6. transferring axis; 7. friction clutch; 8. rubber.

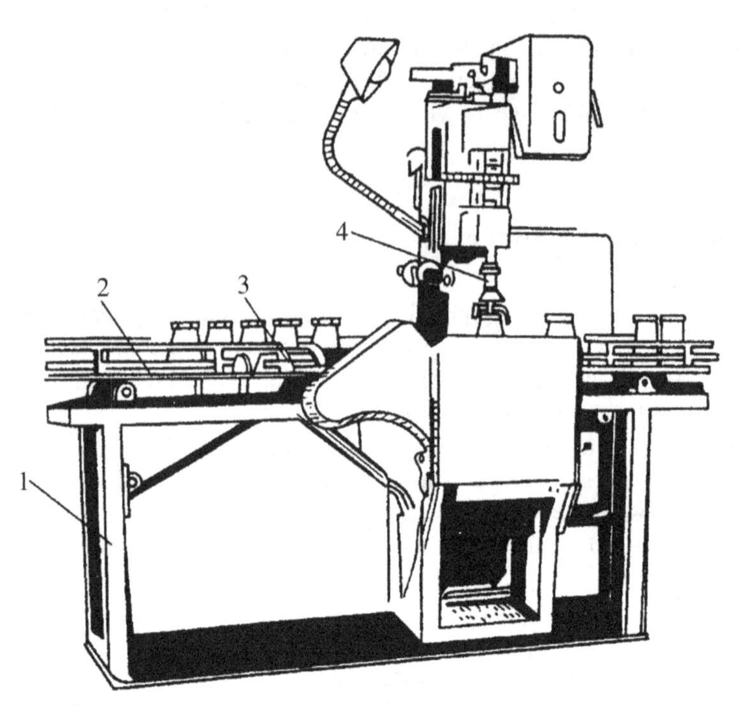

FIGURE 3.43 Outline of the four-screw cap-tightening machine. 1. Mechanical rack; 2. bottle-transporting chain belt; 3. bottle-distributing spiral unit; 4. bottle-holding and cap-tightening system.

the shaft axis and downward movement, the rubber cover holds the cap and seals the container by rotating and screwing the cap. Adjustments of the spring pressure and the friction plate number in the friction clutch vary the screwing cap tightness. Similarly, adjusting the screw on the shaft changes the distance between the shaft and container mouth to meet the requirements of bottles with different heights.

(1) Four-screw cap-tightening machine. This is a single-head intermittent four-screw cap-tightening machine, as shown in Figure 3.43.

The transmission parts are shown in Figure 3.44. The motor drives the worm rod 20 and worm gear 19 via a pulley to make the vertical shaft 17 rotate. The manually capped four-spiral bottles are transferred to the cap-screwing unit and shaft 10 of the working station by the spiral bottle-distributing system and hook chain. The holding unit 3 can then hold the bottles tightly. At the same time, the cloven fan-like wrench 5 on the vertical shaft 10 of the cap-screwing unit places the caps in contact with four-screwing bottles under the conical-pressing sleeve 8 on the vertical shaft. The pressing head 6 on the lower part of vertical shaft 10 presses the cap tightly via the action of lever unit 12. Under the action of sprockets 11 and 16, the vertical shaft 10 allows the wrench 5 to cooperate with the top pressing head and complete the clockwise rotation, thereby tightening the four-screwing-cap bottle. The cap-tightened bottles are then transferred to the next step by the bottle-transferring belt.

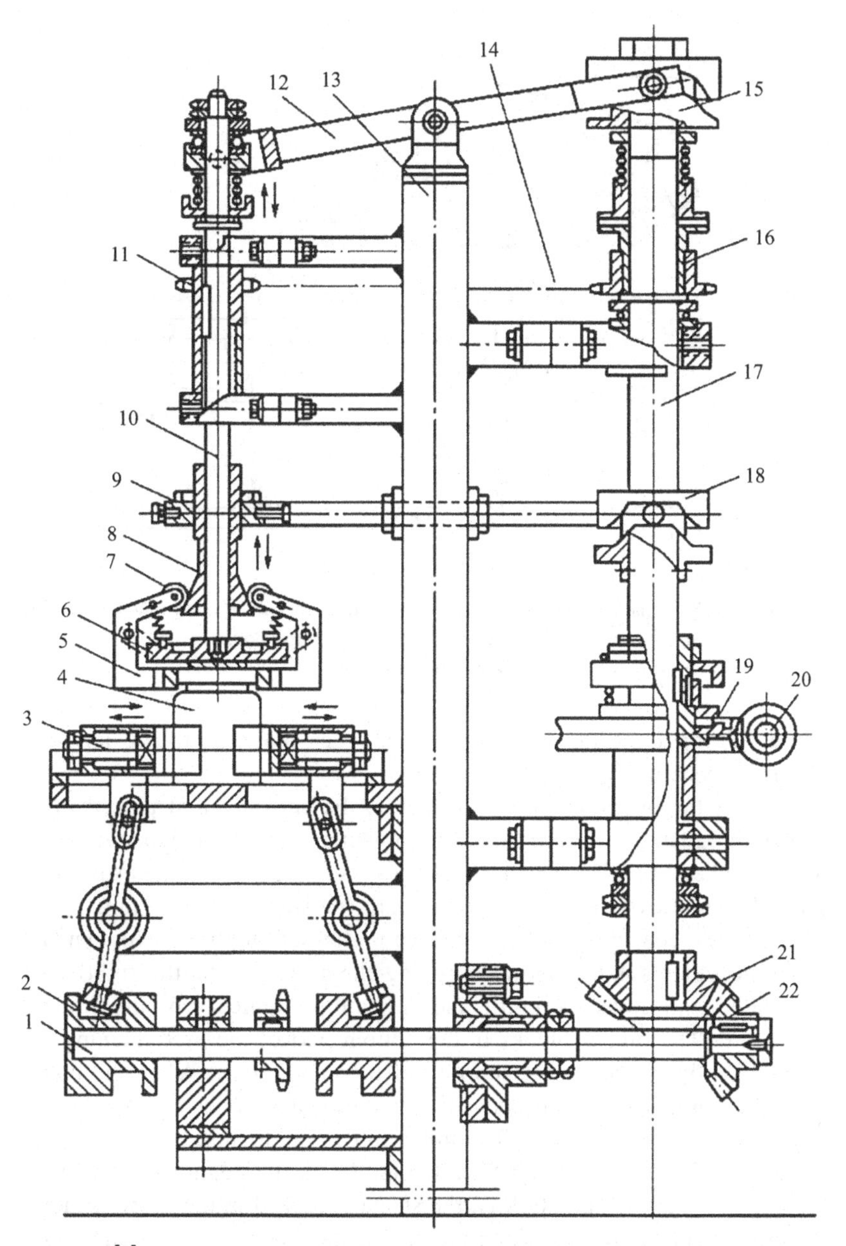

FIGURE 3.44 Schematic diagram of the transmission system of the four-screw cap-tightening machine. 1. Horizontal axis; 2. cam; 3. hand-holding unit; 4. glass bottle; 5. wrench; 6. top pressing head; 7. roller of wrench; 8. pressing head sleeve; 9. fork ring; 10, 17. vertical shaft; 11, 16. sprocket; 12. lever unit; 13. mechanical body; 14. chain; 15. lifting cam; 18. tightening cam; 19. worm wheel; 20. worm rod; 21, 22. bevel gear.

(2) The linear box cap-twisting vacuum-sealing machine is a vacuum-sealing machine, which is suitable for glass bottles with various screw cap shapes. An outline of this machine is shown in Figure 3.45, which consists of the cap-supplying system, prescrewing unit, sealing unit, transmission,

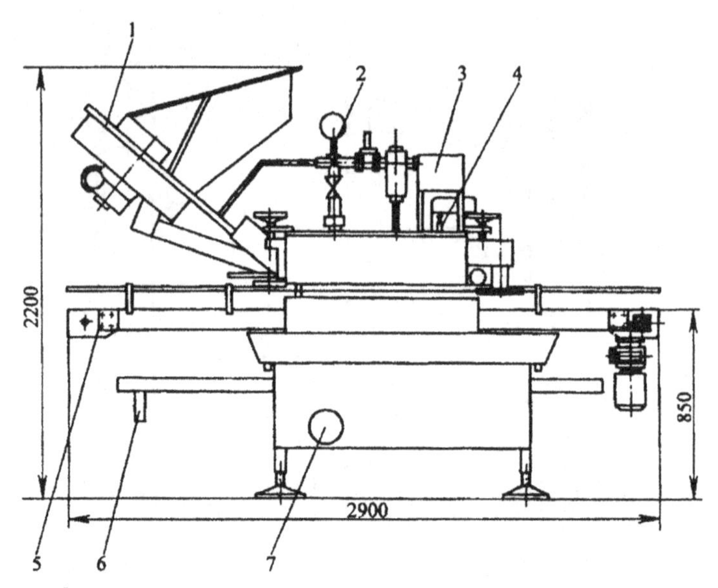

FIGURE 3.45 Outline of the linear box cap-twisting vacuum-sealing machine. 1. Cap-supplying system; 2. steam-piping system; 3. electrical control box; 4. water inlet; 5. bottle-supplying conveyor; 6. water outlet; 7. lifting hand wheel.

bottle-holding unit, steam-piping system, rack-mounted lifting unit, and electrical system.

During the operation of this machine, the caps are first placed into a hopper and then transferred to the cap-organizing unit by the magnet on the rotating wheel. Finally, the caps slide down to reach an inclined suspended state. The bottles filled with materials are transferred to a steaming room by the bottle-conveying belt, while two bottle-holding belts with synchronized movements in the bottle-supplying chain hold the bottles tightly. The suspended caps cover the bottle and they are pretightened by the pretightening unit. The rubber blocks in the bottom part of the pretightening unit are set tightly on the correct portion of the cap surface. When the bottles move forward, the friction between the stationary rubber and the moving cap facilitates pretightening. The filled bottles are then sealed by the sealing unit. Steam expels the air from the bottles in the steam room. After the bottles have been capped tightly, the steam is locked in the empty space in the top part of the bottle. When the bottles have been released and cooled by spraying with cooling water, the steam condenses in the top space of the bottle and forms a vacuum seal.

3.5.4 Vacuum-Packaging Machine

The so-called vacuum-packaging machine is used to place raw materials in airtight plastic bags, before removing the air from the bags and tightly closing the raw materials in the film by sealing the bags. The major advantages of vacuum

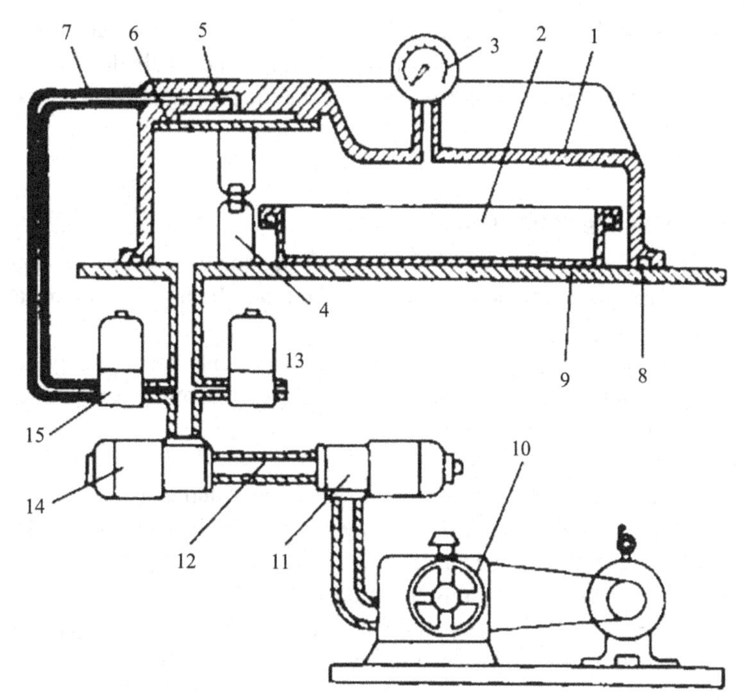

FIGURE 3.46 Principle of the plastic bag vacuum packaging machine. 1. Pressing buckle; 2. material-loading tray; 3. vacuum gauge; 4. heat-sealing head; 5. tightening pressing head; 6. rubber membrane; 7, 12. pipe; 8. sealing ring; 9. working table; 10. vacuum pump; 11, 14. solenoid valve; 13. dual-way solenoid valve; 15. three-way valve.

packaging are the prevention of food oxidation, inhibition of bacterial growth, and extension of the product's shelf-life. In addition, vacuum packaging can prevent bag bulging when the pasteurization process is applied to soft canned products that require heat sterilization.

Figure 3.46 shows the major components and the working principles of a small vacuum packaging machine.

The vacuum pressing cap is usually in a normal open state. Composite plastic bags filled with products are placed into the material-loading tray 2 in the packaging machine. The mouth of the bag is placed on the heat-sealing head and tightened using pressure buckle 1. An O-shaped sealing ring 8 in the bottom of pressing cover is cooperated with the working table 9 to complete the sealing of the bag. At this time, the working room is evacuated using vacuum pump 10 via solenoid valves 11 and 14 and pipes 12. The chamber of rubber film on the pressing head 5 is also evacuated by a vacuum pump using the three-way valve 15. When the indicator on the vacuum gauge three indicates the required level of vacuum, the three-way valve 15 is switched to allow the chamber to connect with the atmosphere via pipe 7. The rubber membrane 6 is driven by the pressure difference between the large chamber and the other chamber, while the pressing head 5 thermally seals the bag placed between the pressing head and the heat-sealing head 4. Next, the solenoid valve 14 is closed, and the dual-way solenoid valve 13 is opened. The main air chamber is subjected to vacuum and the rubber film and the pressing head move upward to complete the whole thermal sealing process. Finally, the pressing cove is opened and the packaged canned products are removed.

3.5.5 Labeling Machine

The labeling machine attaches labels that may include the food name, ingredients, functions, usage method, opening method, and brands on the packaging containers.

Different labeling machines are available for specific packaging purposes, packaging container types, and adhesive label types. The most commonly used labeling machines are as follows.

1. Automatic labeling machine

An automatic labeling machine sticks labels on round cans before packaging, as shown in Figure 3.47. This machine consists of the can delivery device, label-sticking height control device, transmission unit, and mechanical rack.

The working process for this machine is as follows. The cans that need to be labeled roll into the can interval spacer 9 along the can inlet board 8, to ensure that a specific distance interval is maintained between the cans to

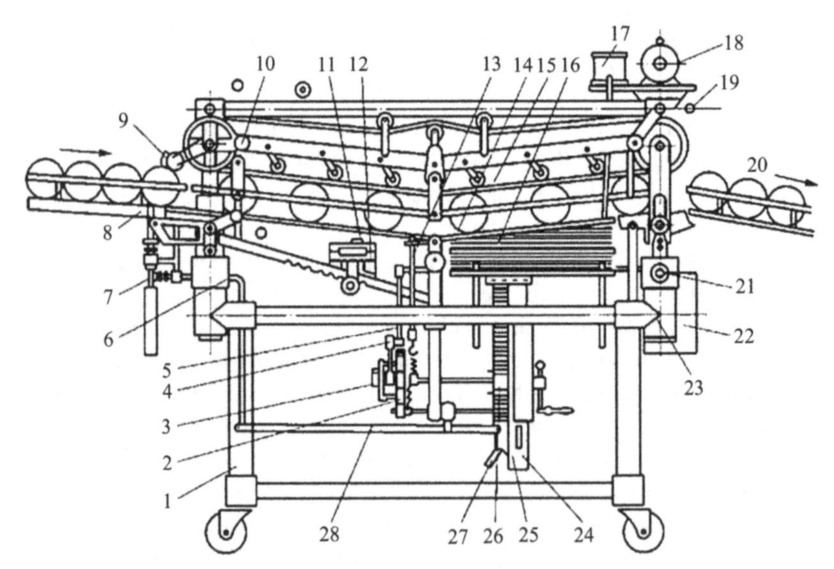

FIGURE 3.47 Automatic labeling machine for round cans. 1. Mechanical rack; 2. ratchet; 3. pawl; 4. pendulum; 5. crank and connecting rod unit; 6. connecting rod; 7. can-inhibiting rod; 8. can inlet board; 9. can interval spacer; 10. hand wheel; 11. small cone; 12. glue box; 13. connecting rod; 14. standard paper height control system; 15. transmission belt; 16. label bracket; 17. glue storage tub; 18. motor; 19. handle; 20. can outlet board; 21. start button; 22. electrical box; 23. glue-containing pressure bar; 24. guide rod; 25. rack; 26. gear; 27. ramp; 28. connecting rod.

avoid collisions or friction during labeling. The cans on the tensioned transmission belt 15 roll forward via friction. There is a glue (or adhesive) box 12 along the route taken by the cans. A small cone 11 is immersed in the glue box. When the can passes through the small cone, its body is covered with glue. As the can rolls forward and reaches the label bracket 16, the label is attached to the body of the can. The can moves forward and the label is then pressed tightly onto the can. Before the label is attached to the body of the can, the end of the label is pressed on by the pressure bar using glue 23. The glue is controlled by the liquid pressure difference in the glue storage tub 17, which controls the amount of glue on the pressure bar. This method facilitates the longitudinal adhesion of the label. The labeled can is then transferred to the outlet board 20 and released through the transferring belt.

During normal operation, the height of the label should be higher than control system 14 height. After labeling, if the label height decreases gradually and is lower than the control block, the connecting rod 13 unit will begin to work and label bracket 16 automatically rises until the label paper height is higher than control system. This procedure is as follows. The can travels through the control system and it is pressed on by the control system if the label paper height is lower than the control system at this time. The connecting rod 13 is raised and the spring is tightened up to make the pawl 3 leave the ratchet 2. Next, the crank connecting rod unit 5 and pendulum 4 pushed the ratchet to complete a fixed axis rotation. The rack 25 under the label bracket mounts the gear chain 26 to drive gear chain upward, thereby raising the label paper bracket until the height of the label paper is higher than the control system. However, the machine will not work if a can passes through the label paper and does not touch the control block. The machine will move intermittently when the label paper is consumed.

The guide rod 24 mounted on the label paper bracket can induce the vertical movement of the bracket. A ramp 27 is located beneath the guide rod. When the label paper is used up, the guide rod rises to make the slanting block touch the connecting rod 28. The connecting rod 28 moves to the left along the ramp 27. Connecting rod 6, which is associated with connecting rod 28, moves to the right under the driving force of the middle lever. The left end of the connecting rod 6 is inserted into the can-inhibiting rod 7. When connecting rod 6 moves to the right, the can-inhibiting rod 7 rises rapidly to the middle of the can-passing channel under the action of the spring, thereby blocking the forward movement of the can and preventing the cans from feeding in the absence of label paper.

This machine is suitable for sticking labels of cans with different diameters and heights. The distance between the top of the mechanical rack and transmission belt can be adjusted using the handle 19. The can height can be adjusted using the hand wheel 10.

2. Portable labeling machine

Figure 3.48 shows a portable labeling machine, which mainly consists of the moveable can and transmission system.

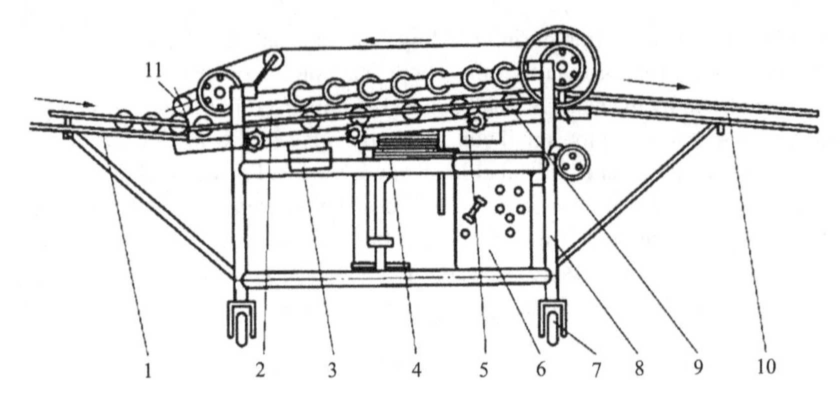

FIGURE 3.48 Portable labeling machine. 1. Inlet board; 2. can-transporting conveyor; 3. glue box; 4. label-supplying bracket; 5. glue pump; 6. electrical box; 7. roller; 8. mechanical rack; 9. can; 10. can outlet board; 11. can divider.

The working process of this machine is as follows. The cans that need to be labeled 9 are brought to the machine by the can-feeding board 1 along the rail. Before entering the conveyer belt, the cans are spaced at a specific distance interval by the can divider 11. The cans are transferred to a rubbing or rolling mode and they move forward via the friction between the conveyor belt 2 and the can 9. When the can first touches the small cones in the glue box 3, the body of the can is covered with glue. The glue-coated can moves to the label paper bracket 4 and the glue on the surface of the can attaches the paper label. As the can moves forward, the attached label is pressed tightly on the can. The labeled cans move to the can outlet board 10 to undergo the next procedure.

When the labeled cans reach the final pressure roller, the pressure roller lifts the cans and drags the glue pump. The glue is then pushed to the glue pipe and it coats the end of the paper label for use on the next can. To enter the next procedure without errors, an electrical unit that automatically excludes cans without labels is installed on the can-releasing board (not shown in the figure).

The can inlet and can outlet boards (1 and 10) can be regulated as required. The rubber rollers 7 installed on the bottom of the machine rack 8 are beneficial for the operation of the machine.

3.5.6 Packing Machine

A packing machine places the packaged cans, bottles, bags, and boxes into corrugated cartons. The packing method is highly dependent on the product shape and requirements. For example, the bottled products should be in an upright position. After pushing against each other, the bottles reach the desired arrangement and are pushed into corrugated cartons in a vertical direction in the top-down or bottom-up modes.

Canned products use a similar packing method to bottled products. In addition, canned products are pushed into a lane separator by horizontal rolling to assume

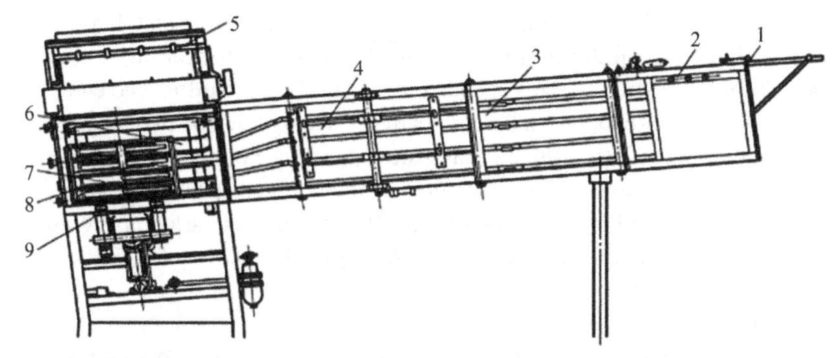

FIGURE 3.49 Packing machine for round cans. 1. Inlet; 2. can-feeding port; 3. storage chamber; 4. quantitative chamber; 5. cardboard inserting unit; 6. main chamber; 7. can pusher; 8. carton; 9. carton-supporting cardboard.

the desired arrangement when opening the carton. The aligned cans lay flat and are pushed laterally into the carton. Moreover, the cans will be pushed into the carton or vertically gripped into the carton if the carton has an opening at one end.

1. Packing machine for round cans

This is a thin tinplate machine used for packing round cans, which works by pushing the flat cans into the lane separator to assume the requisite alignment in cartons with an opening at one end.

This machine consists of can-feeding, can-separating, can storage, and can-pushing systems, as well as a plug-mounted board section, folding table, transmission unit, and pneumatic system. The major components are shown in Figure 3.49.

The operational process consists of the following procedures.

(1) Can feeding. The cans that arrive from the labeling machine reach the inlet 1 for cans along the adjustable track. Depending on the packing requirements, an optical device is installed for counting, and the cans that pass through the can-feeding door will be separated into three or four lanes to fall into the can storage unit.

(2) Can storage. The can storage consists of an adjustable track, storage chamber 3, quantitative chamber 4, and the main warehouse 6 at various points. After separating and blocking, the cans roll into storage chamber 3, which is used as temporary storage depending on the quantitative demand. As the cans pass through quantitative chamber 4, they are allocated on the basis of the quantitative packing specifications (e.g., 24 cans or 12 cans for one batch, i.e., 6×4 cans or 4×3 cans).

(3) Pushing cans into the carton. The empty corrugated cartons are opened manually at one end and placed on the carton set 8, where they are supported by the carton-supporting cardboard 9. The photocell light is occluded by the carton at this time. If the foot switch is depressed, an instruction is sent to the carton to induce a pneumatic system, which activated the gas tank via a microswitch, and the working cycle is repeated to push the cans as required. Finally, the cans are pushed into the carton by the pusher 7.

(4) Inserting the supporting cardboard. To avoid collisions and damage by cans, a supporting cardboard is inserted automatically into the carton to separate each can when a layer of cans is added. A stack of cardboard is erected at the top of the main warehouse 6. The cardboard is usually driven by two-stage gears via the belt of the motor, which installs the supporting cardboard one at a time. After the carton has been filled according to the required specifications, a sign that the carton is full is sent out automatically, and the full carton is laid down by the carton-holding board 9, before it is placed in the raceway for cartons and moved to the next procedure.

2. Carton-generating packing machine

The carton-generating packing machine is a novel piece of packing equipment with a new carton generation function compared with the packing machine described above. Using this machine, the prefolded tubular double-layer corrugated cardboard is stuck and sealed well, and then sent to the machine to produce a carton. After generating the carton, the packing and carton sealing procedures are completed sequentially. This type of machine is suitable for plastic bag products such as orange crystals.

The carton-packing process used by this machine is shown in Figure 3.50. The folded corrugated cardboards stored in the rack 9 are picked up one by one by a sucking disc to generate empty cartons. Next, the empty cartons are sent to the packing position to wait for the plastic film packaged pincushion products that are transferred by the conveyor belt to the product pusher. The products in the bags are sent as one batch to the scraper belt of the packing machine. After the filled bags have been placed in the packing position, the pusher driven by compressed air begins to push the packaged products into the prepared empty carton. The packed corrugated cartons move forward along the conveyor belt. During the movement process, heat-melted resin is sprayed onto the flap of the carton via a resin nozzle. When the cartons reach the pressure position, the flaps on the cartons

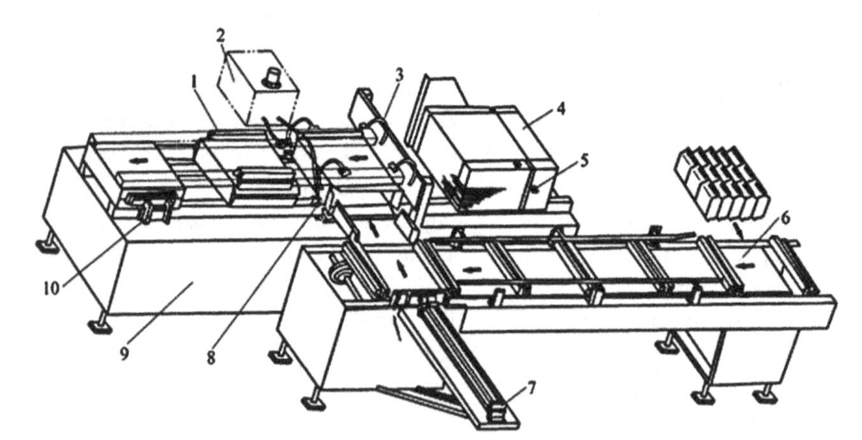

FIGURE 3.50 Carton-generating packing machine. 1, 3. Hinges; 2. heat-melted resin sprayer; 4. corrugated box; 5. sucking disc; 6. scraper belt; 7. filling pusher; 8. nozzle; 9. machine frame; 10. pressurizer.

are subjected to short-term pressure and the carton is sealed well. The adhesive resin is melted by electric heating and it flows into the nozzle to be sprayed onto the flips. The spraying duration is controlled by a valve. All of these procedures are controlled by motor and electric counters, as well as the driving force and controlling system of the machine.

If the bag is too soft or the packaged products in the bag are heavy, the arrangement of the materials should be changed to facilitate horizontal storage rather than upright storage. The arranged products are then pushed into the carton for packaging.

This type of packing machine is suitable for pillow-shaped packaging bags and for the packaging of canned or boxed foods.

During the carton-packing process, additional procedures such as affixed sealing can be added.

This type of machine is characterized by its compact structure, automatic stopping of carton packing driven by a safety propeller that detects the presence of cartons that are open because of glue-sticking failure, alarm signals in the presence of dysfunctional situations, and the reduced consumption of corrugated cardboard via the application of side-opening corrugated carton.

3.5.7 Carton-Sealing Machine

The carton-sealing machine is an automatic pneumatic carton-sealing machine that fixes sealing stickers to the cartons filled with canned products or other products.

As shown in Figure 3.51, this machine consists of the raceway, upgraded cylinder, pace-style conveyor, folding tongue, upper and lower tray racks, upper

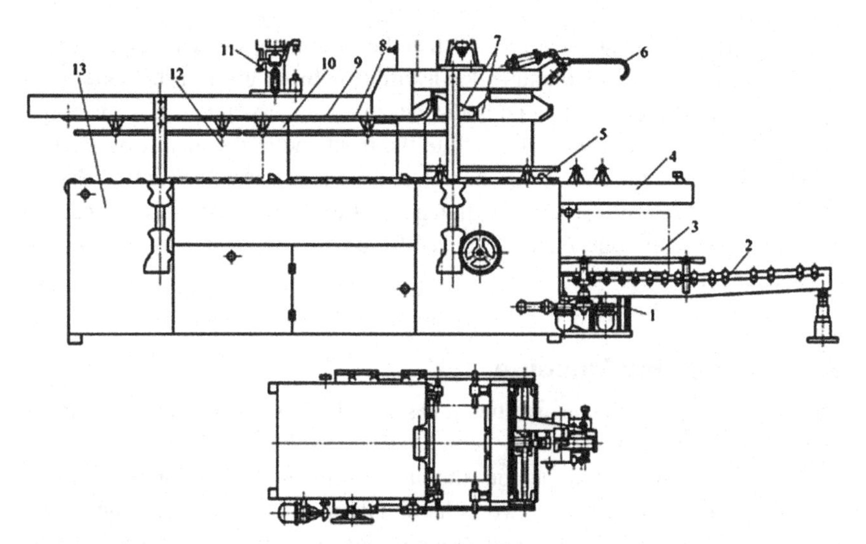

FIGURE 3.51 Carton-sealing machine. 1. Upgraded cylinder; 2. roller; 3. carton; 4. cycle beam; 5. pace-style conveyor; 6. tongue-folding hook; 7. folding tongue plate; 8. paper supply rack; 9. pressure roller; 10. guide rod; 11. paper-cutting device; 12. sealed carton; 13. mechanical body.

and lower water tank, pressure roller, upper and lower cutter, and air-driven system.

The main working process used by this machine is as follows. The opened cartons filled with canned, bagged, and boxed products are transferred to roller 2 of the packing machine via its carton-supporting board. After being pushed by workers, the cartons 3 slide forward along the inclined rollers to touch the process switch. The upgraded cylinder 1 beneath the roller of the packing machine begins to lift the cartons to the top of cycle beam 4 via the pace-style conveyor 5 under the action of the air-driven system. The pace-style conveyor will begin to work after the cartons reach the correct position and the carton arrival signal is transmitted.

The pace conveyor is driven by a two-cylinder pushing rod that allows complete reciprocal movements. It completes a reciprocal movement for each carton. The carton is then pushed to the next procedure by the pawl on the pushing rod. The opened carton is pushed to the rack hook by the pawl on the pace conveyor. First, the small folding tongues in the rear of the carton are tied up by the tongue-folding hook 6. In addition, during the pushing process, the folding tongues in the front of the carton are tied up by the fixed tongue-folder. The large folding tongues are tied up by the folding tongue plates 7 on both sides of the carton and flattened by pressure. The carton is pushed to pressure roller 9 and then pushed to the next step for affixed sealing. The sealing tape mounted on the paper supply rack 8 is passed through the bracket to reach the water-wetting device that wets the tape. The wetted tape is taken to the top of the carton (same as the bottom) and stuck on the carton by pressure roller 9.

During the transportation of the carton by the conveyor pawl, the tape is gradually stuck to the carton, and the carton is sent to the paper-cutting device 11 by the pace-style conveyor pawl. When the carton stops under the paper cutter, the upside cutter cuts the tape downward (the lower edge of the cutter cuts upward). The wheels mounted on both sides of the cutters stick both ends of the tape to the end of the front carton and to the front end of the final carton. After sticking the tape ends on the cartons, the sealing tape forms "⊓" or "⊔" shapes on the carton. The cartons sealed with tape are transferred to the next procedure by the pawl. If self-adhesive tape is used for sealing, the water-wetting device is not necessary.

3.5.8 Strapping Machine

The strapping machine uses various ropes or belts to strap the various sizes of cartons filled with products.

Figure 3.52 shows two specific strapping machines: (a) a strapping machine with the function that straps two belts simultaneously and (b) an automatic strapping machine. The automatic strapping machine can improve the level of automation and it is also equipped with an automatic transmission device and an

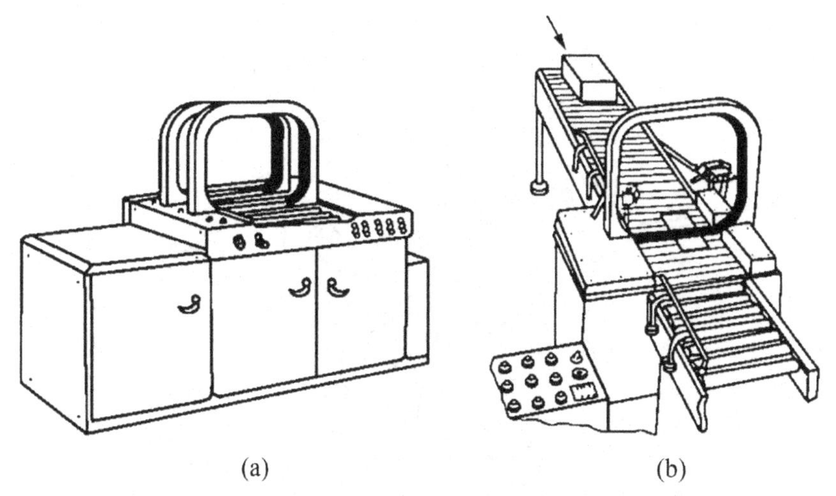

FIGURE 3.52 Two types of strapping machine: (a) semiautomatic strapping machine and (b) automatic strapping machine.

optical positioning device. The cartons that need to be strapped are transferred to the guide rack of the automatic strapping machine by the conveyor belt and the cartons are detected by a photoelectric control device, to trigger strapping machine to activate strapping. The strapped cartons are transported by the conveyor belt.

The most commonly used type is the desktop strapping machine. The strapping job can be completed if the carton is placed on the working table. This type of strapping machine has a wide application range. This machine has a maximum dimension of 600×400 mm for the strapped objects and a strapping speed of 2.5–3 s per object.

During the strapping of cartons for final products, most desktop automatic strapping machines used polypropylene tape (also known as PP tape). Figure 3.53 shows the structure of the PP tape desktop automatic strapping machine. This machine consists of the mechanical rack 2, PP tape roll 1, tape storage box 3, tape-feeding and tensioning device, tape-jointing device, vaulted rail 5, transmission system, and automatic control device.

The working process of the strapping machine is as follows. Plug in the power cord 11, press the start button 6, place the cartons that need strapping on the working table, which has a series of rotatable rollers 12, place the cartons on the vaulted rail 5 to facilitate pushing, and wait for the strapping procedure.

After pressing the foot switch 10, the executive device of the strapping machine begins to operate. The carton passed through the vaulted rail 5 and the PP tape bundles the carton during one cycle. The PP tape is tightened by the belt-tightening device. The closure binder 8 and heater 9 begin to compress the tape ends, as well as welding, cutting, and folding, until the strapping process is complete. The next strapping procedure then commences.

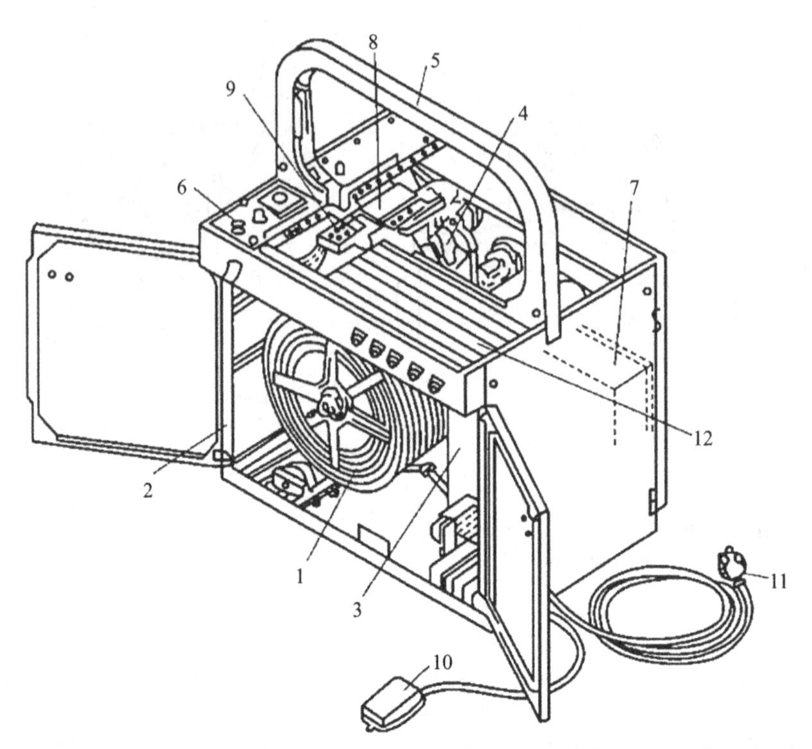

FIGURE 3.53 Schematic diagram showing the structure of the automatic strapping machine. 1. PP tape roll; 2. mechanical rack; 3. tape storage box; 4. tape-feeding roller; 5. vaulted rail; 6. start button; 7. electrical control box; 8. closure binder; 9. heater; 10. foot switch; 11. power plug; 12. roller.

3.6 TYPICAL CANNED CITRUS-PROCESSING PRODUCTION LINE

Production line for canned orange slices in syrup is shown in Figure 3.54.

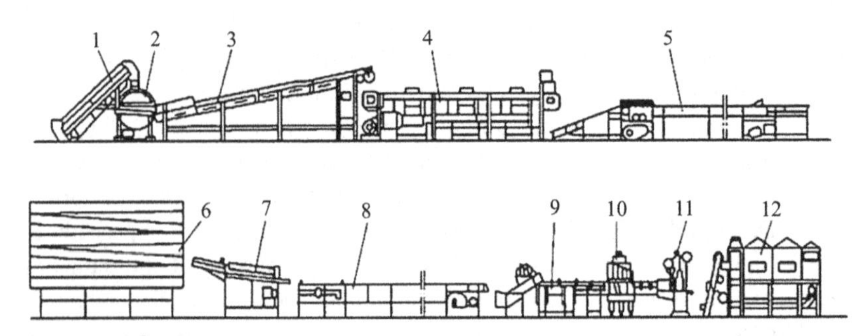

FIGURE 3.54 Production line for canned orange slices in syrup. 1. Scraper elevator machine; 2. hot orange machine; 3. cooling kayaking elevator machine; 4. orange-peeling machine; 5. orange-splitting conveyor; 6. continuous acid–base rinsing tank; 7. orange segment grader; 8. seed-removing conveyor; 9. can-filling and weighing conveyor; 10. juice-filling machine; 11. vacuum-sealing machine; 12. atmospheric continuous sterilization machine.

Chapter | Four

Quality and Safety Control during Citrus Processing

CHAPTER OUTLINE

4.1 Limits and Requirements for Pesticide Residues, Contaminants, and Additives in Citrus and Canned Products According to National Standards 106
4.2 Hazard Analysis and Critical Control Points for Canned Citrus Processing 106
4.2.1 Hazard Analysis and Critical Control Points Overview 106
4.2.2 Application of HACCP in the Production of Canned Citrus Fruits 116
4.3 GMP Control during the Processing of Canned Citrus Products 123
4.3.1 Factory Environment 124
4.3.2 Factory Buildings and Equipment 125
4.3.3 Workshop Isolation 125
4.3.4 Structure of the Factory Buildings 125
4.3.5 Safety Facilities 126
4.3.6 Roof and Ceiling 127
4.3.7 Walls, Doors, and Windows 127
4.3.8 Ground and Drainage 127
4.3.9 Water and Steam Supply Facilities 128
4.3.10 Lighting and Ventilation Facilities 128
4.3.11 Hand-Washing Facilities and Disinfection Pool 129
4.3.12 Locker Room 129
4.3.13 Toilet 130
4.3.14 Warehouse 130
4.3.15 Mechanical Equipment 130
4.3.16 Quality Inspection Equipment 131
4.3.17 Management Mechanism and Personnel 131
4.3.18 Personnel Requirements 132
4.3.19 Education and Training 133
4.3.20 Health Management 133
4.3.21 Quality Management 139
4.3.22 Record Management 142

Y. Shan (Ed): Canned Citrus Processing. http://dx.doi.org/10.1016/B978-0-12-804701-9.00004-6

4.3.23 Tag 142
4.3.24 Establishment and Evaluation of the Management System 142
4.4 Construction of the Traceability Management System for Canned Citrus Products 143
4.4.1 Basic Requirements for System Construction 143
4.4.2 Management Requirements for System Construction 143

4.1 LIMITS AND REQUIREMENTS FOR PESTICIDE RESIDUES, CONTAMINANTS, AND ADDITIVES IN CITRUS AND CANNED PRODUCTS ACCORDING TO NATIONAL STANDARDS

At present, the safety and health quality requirements for various fruits, including citrus and their products, have been increasing at home and abroad. Many developed countries have enacted stringent pesticide residue limits. To ensure that citrus production and business operators know the full standards or requirements for pesticide residues and contaminants in citrus, and the corresponding products in China, this section summarizes the limits and requirements for pesticide residues, contaminants, and additives for citrus and its products. This will provide production and business operators with useful references, so they can ensure that appropriate strategies are adopted during production and business management.

The limits and requirements for pesticide residues, contaminants, and additives have been enacted according to the maximum pesticide residue limits in foods from the GB 2763-2012 National Food Safety Standards, the limits for contaminants in foods from GB 2762-2005, and the application standards for food additives from the GB 2760-2011 National Food Safety Standards. Tables 4.1 and 4.2 show the maximum limits for 107 types of pesticide residues and seven types of contaminants in citrus fruits and their corresponding products in China.

4.2 HAZARD ANALYSIS AND CRITICAL CONTROL POINTS FOR CANNED CITRUS PROCESSING

4.2.1 Hazard Analysis and Critical Control Points Overview

The basic definition of HACCP is hazard analysis and critical control point. HACCP is a preventive food safety technological management system that consists of hazard analysis (HA) and the critical control point (CCP) for foods. HACCP uses the principles and methods of quality control and risk assessment to evaluate the practical and potential hazards and risks in each main food chain, such as raw material planting/farming, processing, distributing, and selling based on food technology, microbiology, chemistry, and physics, which is beneficial for determining the CCPs that have the main effects on the quality of the final products. Thus, appropriate preventative and control strategies can be implemented before the occurrence of risks, which can reduce the minimum level of risks for foods and ensure the higher safety of the final products. Safety,

Table 4.1 Maximum limits for pesticide residues, contaminants, and additives in citrus fruits

	Pesticide name	Citrus fruit (mg/kg)	Resource	Detection method
1	MCPA (sodium)	0.1	GB 2763-2012	GB/T 20769
2	Abamectin	0.02	GB 2763-2012	SN/T 1973, SN/T 2114
3	Paraquat	0.2	GB 2763-2012	SN/T 0340
4	Chlorothalonil	1	GB 2763-2012	NY/T 761, SN/T 0499, GB/T 5009.105
5	Fenthion	0.05	GB 2763-2012	NY/T 761, GB/T 19648
6	Fenbutatin oxide	1	GB 2763-2012	SN/T 0592
7	Benomyl	5	GB 2763-2012	GB/T 23380, NY/T 1680
8	Fenothiocarb	0.5	GB 2763-2012	GB/T 19648
9	Benzoximate	0.3	GB 2763-2012	GB/T 20769
10	Difenoconazole	0.2	GB 2763-2012	SN/T 1975, GB/T 5009.218, GB/T 19648
11	Fenamiphos	0.02	GB 2763-2012	GB/T 5009.145, GB/T 19648
12	Imidacloprid	1	GB 2763-2012	GB/T 23379, GB/T 20769, NY/T 1275
13	Flumioxazin	0.05	GB 2763-2012	GB/T 19648
14	Profenofos	0.2	GB 2763-2012	GB/T 19648, NY/T 761, SN/T 2234
15	Glyphosate	0.5	GB 2763-2012	GB/T 23750, NY/T 1096, SN/T 1923
16	Diflubenzuron	1	GB 2763-2012	GB/T 5009.147, NY/T 1720
17	Kasugamycin	0.1	GB 2763-2012	/
18	Pyridaben	2	GB 2763-2012	GB/T 20769
19	Metriam	3	GB 2763-2012	SN/T 0711
20	Mancozeb	3	GB 2763-2012	SN/T 0711, SN/T 1541
21	Semiamitraz and Semiamitraz chloride	0.5	GB 2763-2012	GB/T 5009.160
22	Phenthoate	1	GB 2763-2012	GB/T 5009.20, GB/T 19648, GB/T 20769
23	Trichlorfon	0.2	GB 2763-2012	GB/T 20769, GB/T 5009.218, NY/T 761
24	Dichlorvos	0.2	GB 2763-2012	NY/T 761, GB/T 19648, GB/T 5009.20
25	Fonofos	0.01	GB 2763-2012	GB/T 19648

Continued...

Table 4.1 Maximum limits for pesticide residues, contaminants, and additives in citrus fruits—continued

	Pesticide name	Citrus fruit (mg/kg)	Resource	Detection method
26	Carbosulfan	1	GB 2763-2012	NY/T 761
27	Diafenthiuron	0.2	GB 2763-2012	/
28	Acetamiprid	0.5	GB 2763-2012	GB/T 23584, GB/T 20769
29	Chlorpyrifos	1	GB 2763-2012	GB/T 20769, GB/T 19648, NY/T 761, SN/T 2158
30	Parathion	0.01	GB 2763-2012	GB/T 5009.145
31	Carbendazim	5	GB 2763-2012	GB/T 20769, GB/T 23380, NY/T 1680, NY/T 1453
32	Famoxadone	1	GB 2763-2012	GB/T 20769
33	Teflubenzuron	0.5	GB 2763-2012	NY/T 1453
34	Flufenoxuron	0.5	GB 2763-2012	NY/T 1720
35	Chlorfluazuron	0.5	GB 2763-2012	GB/T 19648, SN/T 2095
36	Methamidophos	0.05	GB 2763-2012	NY/T 761, GB/T 5009.103
37	Phorate	0.01	GB 2763-2012	GB/T 19648
38	Parathion-methyl	0.02	GB 2763-2012	NY/T 761
39	Phosfolan-methyl	0.03	GB 2763-2012	NY/T 761
40	Isofenphos-methyl	0.01	GB 2763-2012	GB/T 5009.144
41	Fenpropathrin	5	GB 2763-2012	NY/T 761
42	Monocrotophos	0.03	GB 2763-2012	NY/T 761
43	Carbofuran	0.02	GB 2763-2012	NY/T 761
44	Captan	5	GB 2763-2012	GB/T 19468, SN/T 0654
45	Quinalphos	0.5	GB 2763-2012	NY/T 761
46	Dimethoate	2	GB 2763-2012	GB/T 5009.145, GB/T 20769, NY/T 761
47	Bifenthrin	0.05	GB 2763-2012	GB/T 5009.146, NY/T 761, SN/T 1969
48	Phosphamidon	0.05	GB 2763-2012	NY/T 761
49	Phosfolan	0.03	GB 2763-2012	NY/T 761
50	Cadusafos	0.005	GB 2763-2012	SN/T 2147
51	Spirotetramat	1	GB 2763-2012	/
52	Spirodiclofen	0.5	GB 2763-2012	GB/T 19648, GB/T 20769
53	Cyhalothrin and λ-Cyhalothrin	0.2	GB 2763-2012	GB/T 5009.146, NY/T 761

Table 4.1 Maximum limits for pesticide residues, contaminants, and additives in citrus fruits—continued

	Pesticide name	Citrus fruit (mg/kg)	Resource	Detection method
54	Permethrin	2	GB 2763-2012	NY/T 761
55	Cypermethrin and β-Cypermethrin	1	GB 2763-2012	NY/T 761
56	Isazofos	0.01	GB 2763-2012	GB/T 20769
57	Malathion	2	GB 2763-2012	GB/T 19648, GB/T 20769, GB/T 5009.218, NY/T 761
58	Prochloraz and Prochloraz-manganese chloride complex	5	GB 2763-2012	GB/T 19648, NY/T 1456
59	Azoxystrobin	1	GB 2763-2012	GB/T 20769, NY/T 1453, SN/T 1976
60	Methomyl	1	GB 2763-2012	NY/T 761
61	Ethoprophos	0.02	GB 2763-2012	NY/T 761
62	Demeton	0.02	GB 2763-2012	GB/T 20769
63	Fenvalerate and Esfenvalerate	0.2 (Citrus, except oranges and tangerines)	GB 2763-2012	NY/T 761
		1 (Citrus, oranges and tangerines)		
64	Propargite	5	GB 2763-2012	NY/T 1652
65	Thiabendazole	10	GB 2763-2012	GB/T 20769, NY/T 1453, NY/T 1680
66	Hexythiazox	0.5	GB 2763-2012	GB/T 5009.173, GB/T 19648, GB/T 20769
67	Buprofezin	0.5	GB 2763-2012	GB/T 19648, GB/T 20769
68	Dicofol	1	GB 2763-2012	NY/T 761
69	Triazophos	0.2	GB 2763-2012	NY/T 761
70	Triadimefon	1	GB 2763-2012	NY/T 761, GB/T 5009.126
71	Azocyclotin	2	GB 2763-2012	SN/T 0150, SN/T 1990
72	Chlordimeform	0.01	GB 2763-2012	GB/T 20769
73	Triflumuron	0.05	GB 2763-2012	GB/T 20769, NY/T 1720
74	Cartap	3	GB 2763-2012	GB/T 20769
75	Fenitrothion	0.5	GB 2763-2012	GB/T 14553, GB/T 5009.20 (except for bulb vegetables), GB/T 19648, GB/T 20769, NY/T 761

Continued...

Table 4.1 Maximum limits for pesticide residues, contaminants, and additives in citrus fruits—continued

	Pesticide name	Citrus fruit (mg/kg)	Resource	Detection method
76	Methidathion	2	GB 2763-2012	GB/T 14553, GB/T 19648, NY/T 761
77	Amitraz	0.5	GB 2763-2012	GB/T 5009.143, SN/T 0279
78	Isocarbophos	0.02	GB 2763-2012	GB/T 5009.20
79	Clofentezine	0.5	GB 2763-2012	SN/T 1740
80	Terbufos	0.01	GB 2763-2012	NY/T 761, NY/T 1379
81	Aldicarb	0.02	GB 2763-2012	NY/T 761
82	Trifloxystrobin	0.5	GB 2763-2012	GB/T 20769, GB/T 19648
83	Tebuconazole	2	GB 2763-2012	GB/T 19648, GB/T 20769, NY/T 1379
84	Nitenpyram	0.5	GB 2763-2012	GB/T 20769
85	Diniconazole	1	GB 2763-2012	GB/T 19648, GB/T 20769, GB/T 5009.201, SN/T 1114
86	Phoxim	0.05	GB 2763-2012	GB/T 5009.102, GB/T 20769, NY/T 761
87	Bromopropylate	2	GB 2763-2012	GB/T 19648, SN/T 0192
88	Deltamethrin	0.05	GB 2763-2012	NY/T 761
89	Phosmet	5	GB 2763-2012	NY/T 761
90	Imibenconazole	1	GB 2763-2012	/
91	Nicotine	0.2	GB 2763-2012	GB/T 20769, SN/T 2397
92	Omethoate	0.02	GB 2763-2012	NY/T 761, NY/T 1379
93	Etoxazole	0.5	GB 2763-2012	GB/T 19648
94	Acephate	0.5	GB 2763-2012	NY/T 761, GB/T 5009.218
95	Imazalil	5	GB 2763-2012	GB/T 20769
96	Coumaphos	0.05	GB 2763-2012	NY/T 761, GB/T 19648
97	Sulfotep	0.01	GB 2763-2012	NY/T 761, GB/T 19648
98	Fenpyroximate	0.2	GB 2763-2012	GB/T 19648, GB/T 20769, SN/T 1977
99	Aldrin	0.05	GB 2763-2012	NY/T 761, GB/T 5009.19
100	DDT	0.05	GB 2763-2012	NY/T 761, GB/T 5009.19
101	Dieldrin	0.02	GB 2763-2012	NY/T 761, GB/T 5009.19
102	Camphechlor	0.05	GB 2763-2012	YC/T 180

Table 4.1 Maximum limits for pesticide residues, contaminants, and additives in citrus fruits—continued

	Pesticide name	Citrus fruit (mg/kg)	Resource	Detection method
103	HCB	0.05	GB 2763-2012	NY/T 761, GB/T 5009.19
104	Chlordane	0.02	GB 2763-2012	GB/T 5009.19
105	Mirex	0.01	GB 2763-2012	GB/T 5009.19
106	Heptachlor	0.01	GB 2763-2012	NY/T 761, GB/T 5009.19
107	Endrin	0.05	GB 2763-2012	NY/T 761, GB/T 5009.19
108	Lead	0.1	GB 2762-2005	GB 5009.12
109	Cadmium	0.05	GB 2762-2005	GB/T 5009.15
110	Inorganic arsenic	0.05	GB 2762-2005	GB/T 5009.11
111	Chromium	0.5	GB 2762-2005	GB/T 5009.123
112	Fluorine	0.5	GB 2762-2005	GB/T 5009.18
113	Rare earth	0.7	GB 2762-2005	GB/T 5009.94
114	Total mercury	0.01	GB 2762-2005	GB/T 5009.17

GB: National standard of China; GB/T: National recommendatory standards of China.
Note: /, detection method not specified.

Table 4.2 Limits for food additives in canned citrus products

Additive name	Function	Maximum limit (g/kg)	Notes
Saffron yellow	Colorant	0.2	
Disodium stannous citrate	Stabilizer and coagulant	0.3	
Metatartaric acid	Acidity regulator	Appropriate use according to production needs	
Sucralose	Sweetener	0.25	
Acesulfame potassium (acesulfame K)	Sweetener	0.3	
Ascorbic acid and its sodium salt	Antioxidant, color fixative	1.0	Calculation according to ascorbic acid
Carmine and its aluminum precipitate	Colorant	0.1	Calculation according to carmine
Sodium cyclamate, calcium cyclamate	Sweetener	0.65	Calculation according to cyclohexyl sulfamic acid
Calcium chloride	Function stabilizer, coagulants and thickener	1.0	

wholesomeness, and quality underpin the preventative management system used to eliminate hazards and ensure food safety. This system provides guidance to improve the supervision of food hygiene inspectors, as well as allowing food manufacturers to guarantee the product quality and enhance product quality management during merchandise competition.

The United States was the first country to propose the principles of HACCP and to implement HACCP control systems during food processing. In 1960s, the Pillsbury Company and National Aeronautics and Space Administration in the United States cooperatively developed aerospace foods, when Dr Howard Bauman proposed the HACCP concept to ensure that no pathogens or toxins are present in foods. In 1973, the US Food and Drug Administration (FDA) successfully applied HACCP to low-acidic canned food processing. The federal government has also applied HACCP principles and methods to the processing of aquatic products, meat products, fruit and vegetable juices, and other foods. The FDA has enacted a series of HACCP regulations to monitor the safety of food production and food processing enterprises, as well as the foods that are imported into the United States. The application of HACCP has been encouraged by the World Health Organization (WHO) and the International Commission on Food Microbiology.

The Food Hygiene Committee has developed a standardized approach to the application of HACCP for all member states. The Codex Alimentarius Commission (CAC) also strongly encourages the application of HACCP systems in the global food industry. In 1997, CAC developed an HACCP system and its application guidelines. According to these guidelines, HACCP can be applied to the whole food chain from the original producers to the final consumers. The application of this system will be helpful for developing the regulations for inspection authorities, thereby improving the credibility of food safety in international trade. At present, the HACCP system has become publicly recognized as an effective food safety control system worldwide. The HACCP concept was introduced into China in the 1980s. In 1990, the Food Science and Technology Committee of the former State Import and Export Commodity Inspection Bureau conducted research into HACCP and developed an HACCP quality management system. At present, the application of HACCP to the Chinese agriculture or food processing industry has become a scientific and technological project for food safety as part of the "10th 5-year" Plan.

To ensure the implementation of HACCP management systems, we must first fully understand the concept and significance. We must also understand the basic principles of the HACCP system so that it can be put into action more comprehensively. Detailed descriptions of the seven basic principles and brief explanations of the HACCP system are as follows.

Principle 1. HA and control strategies during processing. A list has been produced of the potential risk processes or steps during processing, and corresponding control strategies have been described.

Principle 2. Decision of CCP. The CCPs have been determined to implement correct control strategies that facilitate hazard prevention, hazard elimination, or hazard reduction to an acceptable level, such as heating, refrigeration, and specific sterilization procedures should be noted and controlled at each significant hazard point. CCP cannot include all points of the inputs or output for each significant hazard, step, or process, but the CCP decision points are beneficial for the establishment of a CPP decision tree.

Principle 3. Determining CCP limits (CL). This principle demands that the corresponding preventative measures for CCP must meet the specific requirements. For example, high or low temperatures, long or short control durations, pH ranges, and salt concentration should be within the appropriate limits. Control limits are important for ensuring food safety. Each CPP has one or more CL values. If the CL value deviates during the operational process, appropriate corrective measures should be implemented to ensure the safety of foods.

Principle 4. Establishing a CCP monitoring program. This principle demands a series of planned observation and measurements, such as the temperature, time, pH, and water content, to evaluate whether CCP is within the control limits, as well as accurately recording the monitoring results for future verification and identification. In addition, the monitoring personnel should be clear that their duty is to control all the important components of CCP. The monitoring personnel must report risky processes and record the processes or products that do not meet the requirements of CCP and take corrective action immediately. All records and documents related to CCP should have the supervisor's signature.

Principle 5. Establishing corrective measures. If the monitoring results indicate that the process is out of control, appropriate corrective measures should be implemented immediately to reduce or eliminate the potential hazards caused by uncontrolled processes, thereby restoring the uncontrolled processes back to their normal conditions. Corrective measures should be planned during the development of HACCP systems, where their functions include (a) determining whether the foods produced in the uncontrolled state need to be eliminated, (b) correcting or eliminating the uncontrolled causes, and (c) recording the performance of corrective measures.

Principle 6. Establishing an effective record-keeping system. The required records include (a) the HACCP plan purpose and scope, (b) product description and identification, (c) flow chart of the process, (d) HA, (e) auditing tables for HACCP, (f) basis of determining the CLs, (g) validation of CLs, (h) monitoring records including deviations from CLs, (i) corrective measures, (j) records related to verification activities, (k) inspection records, (l) cleaning records, (m) product labels and traceability, (n) pest control, (o) training records, (p) records of approved vendors, (q) product recycling records, (r) auditing records, and (s) modifications, reviews, and corresponding documents for the HACCP system. During the actual management process, records provide an important control strategy, which facilitate the adjustment of processing or prevent uncontrolled CPP. Therefore, records are important components of the successful implementation of HACCP plans.

Principle 7. Establishment of procedures to ensure the correct operation of HACCP systems. Even if HA, CCP monitoring, corrective measures, and effective record maintenance procedures are implemented, food safety cannot be guaranteed solely by the establishment and operation of an HACCP system. Thus, the critical requirements still include (a) verifying that each CCP is enforced strictly according to the HACCP plan, (b) confirming the completeness and effectiveness of the overall HACCP plan, and (c) validating that the HACCP system facilitates normal and efficient operating conditions. These three elements compose the HACCP verification procedures.

The analysis of potential hazards, identification of CCP during processing, and establishment of CLs are three important steps during the risk assessment of food and the execution of an HACCP program. These critical steps should be assessed by technical experts, while the other steps should be addressed by quality management experts.

HACCP is a real logic control and evaluation system that ensures the prevention of foodborne diseases and the supply of healthy and safe foods, thereby improving the satisfaction of consumers. In recent years, HACCP has been applied increasingly extensively, where it is characterized by simple operation and rational implementation compared with other quality control systems.

i. Comprehensiveness: HACCP is a systematic approach that encompasses all aspects of food safety (from the raw materials, planting, and harvesting to the purchase and use of the best products by consumers) that can identify and predict potential hazards.
ii. Prevention as the focus: HACCP is a preventive quality assurance system that can prevent hazards from entering food processing, rather than retrospective testing of the final products.
iii. Improving product quality: HACCP is a product quality control system that ensures higher product quality and food safety for consumers.
iv. Excellent economic benefits: HACCP is a preventative quality control system that reduces losses, product costs, and working intensity by frontline workers, as well as improving their work efficiency.
v. HACCP affirms the basic responsibility of the food industry in ensuring food production quality and safety. The primary responsibility for food safety belongs to food manufacturers or distributors.
vi. HACCP can improve the efficiency of government supervision and management. The official examinations can be focused on the most hazard-prone areas by examining HACCP monitoring records and corrective strategies to understand all of the conditions in food factories.
vii. HACCP provides a link between the food industry system and the management system from government, which is beneficial in improving the relationships among factories, official management, and consumers.

HACCP systems can be used to reduce the hazards or risks of food safety, but it is not a zero-risk system. Thus, HACCP systems can supplement other quality management systems. The combined application of HACCP system and other quality management systems can result in greater synergistic benefits.

The implementation of HACCP is very complex. According to the basic principles described above, the CCPs in the overall operation process must be found initially. The establishment of CCPs is fundamental for the implementation of HACCP control systems. Food processing monitoring must begin with the cultivation and harvesting of raw materials and continue until the consumption of products by the consumers. The CCP decision tree is as follows:

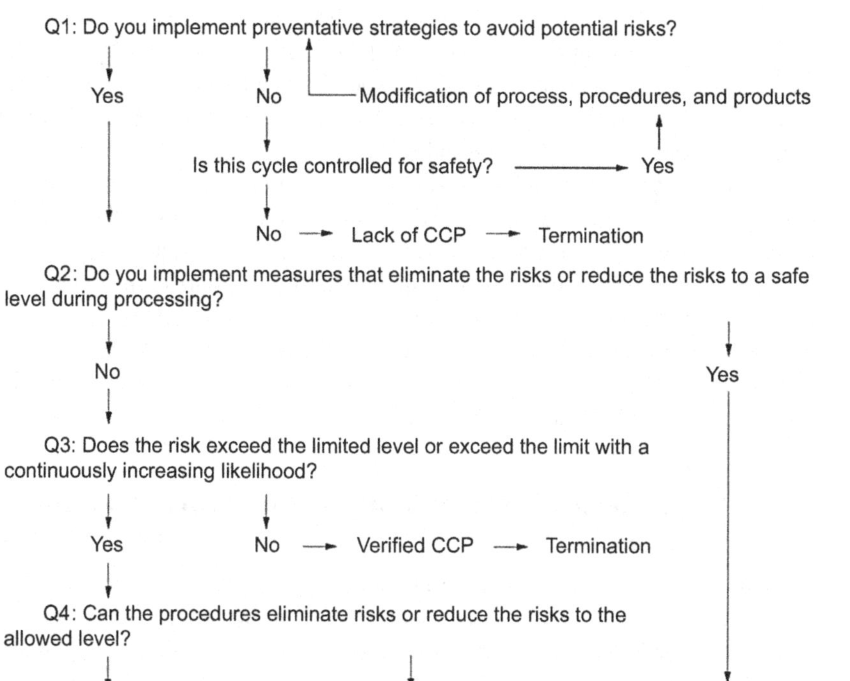

Different industries, manufacturers, products, and producers have different CCPs. Even in the same operational step, the CCPs vary for different risks. The CCP decisions should be based on actual situations and implemented according to the local conditions, rather than the simple and direct replication of existing systems.

Food is a fundamental material for human survival and reproduction. The major issues related to foods at present are as follows: (a) microbial contamination of foods is variable, (b) chemical contamination of foods is increasing, (c) poor diets and bad lifestyles are novel safety factors, and (d) new types of foods are safety issues, such as genetically modified foods, fortified foods, and functional foods. Thus, it is highly desirable to establish and improve food safety control systems as soon as possible. In the current context, food safety issues extend beyond the scope of traditional food hygiene or food contamination, and they involve the protection and management of the whole food chain. Therefore, these issues require combined efforts by governments, scientists, entrepreneurs, and consumers. These ideas are encompassed in the principles of HACCP, which is the most promising food safety control system in the twenty-first century. The HACCP system has a short history. Thus, there is little specific operation experience and insufficient basic research and its development and application will be a long and arduous task. However, as a novel control model, management control using the HACCP system can eradicate unqualified products during the production process. Controlling biological, chemical, and physical contaminants effectively, improving food safety, reducing the risks during the production and sale processes, and promoting the management-level improvement of food production enterprises will provide an effective, systematic, and rational control model. The application of HACCP to the food industry improves food safety, and it enhances the company's reputation in the community, especially export companies, by ensuring the product quality, establishing an excellent brand, and participating in international competition. Excellent monitoring and control mechanisms, and the completion of records, will eventually become an effective demonstration of the manufacturer's responsibility. At the same time, these efforts can also satisfy the requirements for appropriate supervision by all administrative and operational departments, as well as accurately addressing consumer complaints and other issues, which will be conducive for the smooth implementation of food safety process in China. Thus, we have sufficient reason to expect that HACCP will develop into an effective food safety control system.

4.2.2 Application of HACCP in the Production of Canned Citrus Fruits

Since the success of the Apollo program in this century, the role of the HACCP system in food safety and product quality has been recognized. With China's accession to the World Trade Organization as one of the food safety and quality control strategies, application of the scientific HACCP management system in the production of canned citrus products is essential.

Canned citrus products are not only sold in the domestic market but also exported to Europe, Southeast Asia, and other countries. To ensure product quality, the establishment of an HACCP safety system to prevent hazards is desired.

1. Hazard control during the production of canned citrus products

1) Technological process

Examination and acceptance of citrus fruits→classification→upgrading→blanching→peeling and splitting→acid and alkali treatment→rinsing→sorting and selecting→canning→weighing→exhausting→sealing→sterilization→spraying and cooling→drying→coding→metal detecting→storage→checking and packing→final products.

2) Worksheet of HA during the production of canned citrus products

According to the above process, biological, chemical, and physical hazards during the production process of canned citrus products have been comprehensively analyzed, and a series of hazard prevention measures have been proposed, as shown in Table 4.3.

3) HACCP plan for the production of canned citrus products, as shown in Table 4.4.

2. Hazard analysis for the production of canned citrus products

Table 4.3 Worksheet of hazard analysis during the production of canned citrus products

Processing step	Hazard analysis	Is the potential hazard significant?	Control points	Prevention strategies	Conform to CCP
Examination and acceptance of cans and caps	B: Pathogen contamination	Yes	Bacterial contamination again during double crimping	Control the second crimping and isolate it, then control the quality of the pouring glue	Yes
Examination and acceptance of raw materials	C: Pesticide residues (organic phosphine), heavy metals (copper, aluminum, arsenic), and other foreign matter B: Citrus decay, serious pest diseases P: Impurities	Yes	Excessive amount of pesticides used in the growth of raw materials; excessive amount of copper, aluminum, and arsenic in soil and water; citrus decay, pathogens, and parasites on the surface of raw materials; possible presence of metal and glass fragments	Receive the fruits according to the compliance certification on pesticide residues, heavy metals census of raw material, control the rot rate below 5%, fruits with worms <2% and immediately discharge the impurities	Yes
Blanching	If blanching is incomplete, it will affect the next process. If blanching is excessive, it will affect the taste of the final products	No	Strictly control blanching time and temperature. Blanching temperature should be between 95 and 100 °C and blanching duration should be 25–45 s	Use thermocouple thermometer to measure, and inspect by peeling	No
Peeling and splitting	If peeling is incomplete, the next process will be affected	No	Seriously and carefully conduct the procedure.	Require the staff to carefully conduct the procedure	No
Acid and alkali treatment	If acid and alkali treatment is incomplete, the product quality will be affected and the sackcloth will not be completely removed completely. If excessive acid and alkali treatment was performed, the appearance and taste of the final products will be soft and rotten	No	Strictly control the concentration of acid and alkali as well as the treatment time. The concentration of HCl acid is 0.6–0.8%, and the processing time is 40 min. The concentration of NaOH should be 0.6–0.8% and the processing time should be 65 min. The temperature of NaOH should be within the temperature range of 40–44 °C	Carefully prepare the concentration of acid/alkali, measure the temperature of lye (if the temperature is too high, the citrus fruits will have a cooked flavor, and its flesh with be soft and rotten. If the acid/alkali concentration is too low, the sackcloth will be difficult to remove), calculate the treatment time of the acid/alkali liquid	No
Rinsing, sorting, and selecting	If rinsing is incomplete, the final products will have a smell. If sorting and selecting are not good, the shaping rate and the quality of the final products will be affected	No	Rinse thoroughly until no smell (acid and alkali taste) is detected. Sort and select according to the consistency of the segment shape	Require the staff to carefully perform this process	No

Continued...

Table 4.3 Worksheet of hazard analysis during the production of canned citrus products—continued

Processing step	Hazard analysis	Is the potential hazard significant?	Control points	Prevention strategies	Conform to CCP
Weighing and canning	Excessive loading will affect heat penetration; less loading will affect the quality of final products. Uniform loading should be conducted as often as possible	No	Do not leave small segments on the edge when canning to avoid affecting the sealing process and leave sufficient headspace of 3–8 mm on average	Consistently fill the cans, accurately conduct weighing by using a tray and repeating the process to verify the weight. There should be tolerance for generating canned products of substandard quality	No
Preparation of liquid	Both too high and too low sugar concentrations will affect the taste and shelf-life of the final products	No	Avoid preparing sugar liquid using iron utensils. The sugar liquid must be used at once and filtration is required after preparation. The concentration depends on the content of soluble solids before filling and the concentration of the liquid in the final products after opening	Perform the procedure according to the operation standard	No
Filling and exhausting	Low temperature in the center of the can will lead to inadequate vacuum	No	Control the temperature of the soup, inject hot sugar liquid to improve the initial temperature of the cans, which can enhance the bactericidal effect	Measure the temperature using a thermometer	No
Sealing	B: Poor sealing will cause leakage and microbial growth inside the cans	Yes	Seal immediately after exhausting to avoid decreasing the internal vacuum as the temperature drops, the iterative rate is ≥60%, the tightness is ≥60%, and the rate of full joint cover is ≥60%	Designate a person who shall be responsible for this job. Cans that show signs of improper sealing must be removed and resealed. Periodic sampling and measuring the curling using instruments are required. If the measured difference in size is within the permissible limits of the operation, then adjustment can be performed during the machine break; if the difference exceeds the limit, stop immediately and repair the machines	Yes

Sterilization	B: Incomplete sterilization will increase the risk of emergence of pathogen residuals	Yes	Strictly enforce the temperature and time of sterilization. The sterilization formula is: 8-12-12 min/125 °C (large cans) or 8-12-12 min/12 °C. The three steps of sterilization are exhausting and heating, sterilizing, and anti-pressure cooling	Measure the temperature with a metal thermometer and monitor; carefully conduct the operate sterilization steps	Yes
Spraying and cooling	Leakage will lead to the breeding of pathogens. Microbes in the cooling water should proliferate any further	No	Foods will contract during cooling because the temperature drops and the empty space partly forms the vacuum state. The filler inside the crimping sealing lid will soften due to the high temperature during sterilization. Some cold water will be sucked into the cans during the cooling process. The microbes will also be drawn into the cans, which could later cause putridity	Use chlorine to measure the reagents. It is preferable to disinfect cooling water with 0.0035% chlorine solution	No
Checking and packing	Rust corrosion, bulging cans, and flat cans mixing with final products will result in unnecessary loss	No	There shall be no rusty cans, bulging cans, and cans with inadequate vacuum mixed with the final products	Manually check the external features of the cans	No

B: Biohazard; C: Chemical hazard; P: Physical hazard.

Table 4.4 HACCP plan for the production of canned citrus products

CCP items	Significant hazard	Critical limits of prevention strategies	Monitoring				Corrective action	Record	Verification
			What	How	Frequency	Who			
Examination and acceptance of empty cans and caps	Poor double crimping will cause contamination and microbe proliferation	Sealing: Tightness >60% Iterative rate >60% Integration rate >60% Poured glue: Ammonia glue 70–110 mg, Grees glue, 40–80 mg	Double crimping	Anatomically examine by using a caliper or a projector	Once for each batch	Quality control staff	**1.** Reject any cans and covers provided by nonqualified suppliers **2.** Reject any cans and covers that do not have inspection certificates **3.** Reject any unqualified cans and covers	Review the record of double crimping cans, check the quality of the glue	Verification record
Examination and acceptance of raw materials	Pesticide residues, heavy metals, mycotoxins	Compliance certification of pesticide residues and heavy metals census; the number of rotten fruits is <5%, and the number of pest fruits is <2%	Exceeded pesticide residues and heavy metals, rotten fruit, and pests	**1.** Compliance certification **2.** Control the number of rotten fruits	Each cart	Quality inspector	**1.** Reject any raw materials that do not have a compliance certification of pesticide residues and heavy metals census **2.** Reject any raw materials that have exceeded the number of rotten fruits and pests	Record of the quality of raw materials	Check for excess pesticide residues and heavy metals in the semifinal products and final products everyday
Sealing	Poor crimping will cause secondary contamination and microbial proliferation	Sealing; tightness >60%, iterative rate >60%, and integration rate >60%	Double crimping	Anatomically examine the quality of sealing by using a caliper or a projector	Anatomically examine once/3 h, visually a check once/1 h	Inspector	Recheck the products. If the products qualify, then maintain normal production; if the products fail, then check the machines and detain all products from the last inspection to this time, and resolve the issues after accreditation	Record of the inspection of double crimping cans	Verification record
Sterilization	Incomplete sterilization will result in the emergence of residual pathogens	8-12-12 min/125 °C (large cans), 8-12-12 min/121 °C (small cans)	Sterilization temperature and time	Monitoring	Each jar	Operator	If deviations occur, correct these discrepancies according to the corrective procedures	Sterilization records, corrective records of cans	Verification record

1) Hazards caused by physical factors

The production of canned citrus products is affected by impurities in raw materials, incomplete removal of the sac coating, and residual metal shavings during the sealing process, as well as other physical factors. Therefore, the efficient removal of physical impurities during raw material examination and acceptance, peeling, sac-coating removal, and sealing is required.

2) Hazards caused by chemical factors

It is necessary to verify the original production place and detect pesticide residues, heavy metals, toxic chemicals contaminating raw materials, and water pollutants. These substances must be strictly controlled; otherwise, they could harm human health.

3) Hazards caused by microorganisms

(1) Raw materials.

The selection of raw materials must ensure the absence of pollution, pesticides, or heavy metal residues. The incidence of rotten and pest-laden fruits must be strictly controlled to less than the established standards in order to ensure the control of microbial contamination from all possible sources.

(2) Process.

The process design must be smooth. The long-term accumulation of raw materials must be avoided to prevent the emergence and growth of microorganisms. If the soup in the can is not sufficiently heat, this may facilitate bacterial proliferation. Poor sealing of cans can also easily lead to secondary bacterial contamination of the canned products. Incomplete sterilization could also lead to the emergence of residual pathogens.

(3) Hygiene.

Incomplete sterilization of tools and poor hygienic activities of the production staff could result in microbial contamination. Therefore, three steps of hygiene control for the workshop such as preproduction, production, and postproduction are necessary. Prior to production, environmental hygiene standards should be checked to determine whether they fulfill the established standards, including disinfection of hands, feet, and overalls of the staff. Production should not start prior to disinfection of all industrial equipment using 200 mg/kg calcium hypochlorite. During production, raw materials, empty cans, water, overalls, hands of operators, tools, and equipment must be examined according to microbiological criteria. If any of these components fail the disinfection examination, recleaning and redisinfection are required. After production, no staff member is allowed to leave until the tools and equipment are cleaned and disinfected. These procedures will ensure the hygiene of the workshop and the quality of the final products.

3. CCP for the production of canned citrus products

1) Examination and acceptance of empty cans and covers

A person who shall be responsible for the quality of double crimping of empty cans and the quality of covers should be identified. The person must fully enforce the operation procedures, prepare detailed records, and correct any problems according to the corrective procedures.

2) Examination and acceptance of raw materials

First, the production place, pesticide residues, and heavy metal residues should be examined. The examination and acceptance of raw materials should be performed according to the established procedures of purchasing raw materials and the corresponding contracts. Any unqualified raw materials should be rejected.

3) Sealing

The process and workshop inspectors are in charge of the critical inspection according to the required procedures. They must conduct major checks, prepare detailed records, identify problems, and take corrective actions to ensure the quality of the sealing.

4) Sterilization

A person(s) who shall be in charge of the sterilization process according to the prescribed sterilization formula and shall control the important parameters such as sterilization time and temperature should be identified. Rules must be strictly followed to ensure compliance with sterilization requirements.

4) Control limits of CCP

(1) Examination and acceptance of empty cans and covers: See Table 4.4.

(2) Examination and acceptance of raw materials: The examination and acceptance of raw materials must be performed according to the purchase contracts or the inspection procedures of purchased goods.

(3) Sealing: See Table 4.4.

(4) Sterilization: See Table 4.4.

5. Monitoring records

The HACCP system involves a series of records. The major records during the production process of canned citrus products include the HACCP standard plans and supportive documents for designing plans such as raw material checking records, sampling records, weighing records, sealing records, sterilization records, semifinished product inspection records, hygiene inspection records, correcting records, final product inspection records, and other verification activity records. HACCP monitoring records are generally reserved as 5a in strict accordance with the recording control procedure.

To ensure the smooth execution of HACCP plans during the production process of canned citrus products, good manufacturing practice (GMP) and sanitation standard operating procedures and the corresponding supportive documents must be formulated. In the quality control department, a person should be in charge of daily inspections of the raw materials, production process control, monitoring records, corrective measures, environmental hygiene, workshop hygiene, storage, transportation, and sales. The workshop inspector and the quality control department must inspect HACCP monitoring records on a daily basis. The HACCP group must regularly inspect the operating situation of HACCP. Once the situation is out of control, corresponding departments must take corrective measures in time, and the HACCP group will reinspect the situation to ensure that the normal operation of the HACCP system is established.

6. Conclusions

During the production of canned citrus products, good or bad implementation of HACCP is highly correlated with the implementation of GMP, the establishment of a good quality management system in the enterprise, the advanced detection technology for hazardous substances, food pollution detection and control technology, and food safety control technology.

HACCP is applied to the production of canned citrus products to determine the potential hazards in every aspect of the production process to identify CCPs that could impact product quality and to develop appropriate preventive measures for each CCP. Establishment of a secure monitoring system will minimize potential harmful factors in the production process as far as possible, which will not only improve the safety of canned citrus products but also ensure the quality of the final products. It will not only improve the credibility of the enterprise but also increase the level of competitiveness of the products in the market.

4.3 GMP CONTROL DURING THE PROCESSING OF CANNED CITRUS PRODUCTS

GMP is an autonomic management system that focuses on the monitoring of product quality and the implementation of hygiene and safety control during the production process. It is a set of mandatory standards applicable to pharmaceutical, food, and other industries that require the companies to achieve health quality requirements according to the national laws and regulations with respect to raw materials, personnel, facilities, equipment, production process, packaging, transportation, and quality control. GMP aims at forming operational specifications to help companies improve the hygienic standards of their work environment and to discover and resolve problems involving the production process.

Decades of application have proven that GMP is an effective tool for ensuring product quality. Therefore, the UN Codex Alimentarius Commission (CAC) designated GMP as an essential procedure that should be integrated into the HACCP principles. In 1969, the WHO recommended the use of GMP around the world. In 1972, the 14 member states of EC announced the general rules of GMP. In 1975, Japan initiated the development of various food hygiene norms. The application of GMP in the food industry in China began in the 1980s. In 1984, to strengthen the supervision and management of food production enterprises and to ensure the safety and healthy quality of exported foods, the former State Administration of Commodity Inspection formulated the minimum hygiene requirements for export food factories and warehouses. This provision is similar to GMP hygiene regulations. In 1994, the Ministry of Health revised this to the "Hygiene Requirements in Export Food Factories and Warehouses." In 1994, the Ministry of Health developed the "General Hygiene Norms in Food Enterprises" (GB 14881-1994) as the national standard, with reference to the "General Principles of Food Hygiene" of

FAO/WHO CAC/RCP Rev.2-1985. Later, it released the "Hygiene Specifications of Canned Food Factories" and the "Hygiene Norms of Liquor Factory," along with 19 other national standards.

Although these national standards are mandatory, these are not extensively implemented because of several limitations, the lag of standardization in China, the backward state of hygiene conditions and facilities in food enterprises, as well as the poor dissemination and monitoring of measures by relevant government departments. Therefore, the Ministry of Health decided to develop GMP for food manufacturing enterprises on the basis of the revisions of the original hygienic regulations. In 2001, through the joint efforts of the health sectors of Guangdong, Shanghai, Beijing, and Hainan, together with a number of enterprises, the Ministry of Health established a cooperative group that was mandated to develop GMP for five food products, including dairy, delicatessen, confitures, drinks, and probiotics food. This group established the general principles, basic format, and corresponding contents of GMP, which not only enhances the operability and scientific soundness of the policy but also increases and specifies GMP contents and proposes requirements for good production equipment, a reasonable production process, sophisticated quality management, and a strict testing system.

GMP for citrus processing enterprises is as follows.

4.3.1 Factory Environment

Factories should not be located in areas that are exposed to pollution. There should be no dust, harmful gas, radioactive substances, and other diffused sources of pollution around the factories. There should be no potential breeding places for insects. Otherwise, strict food contamination prevention measures should be taken. The surroundings should be easy to clean, and there should be no waterlogging, mud, and dirt on the ground. The factory space should be built of concrete, asphalt, or green plants. Roads near and within the factory complex should be paved with concrete, asphalt, or other hard materials to prevent the accumulation of dust and water. There should be no bad odor, harmful (toxic) gas, soot, or other health hazard facilities in the factory site.

Animals that have no role in the production process should not be kept within the factory area. The feeding area of experimental animals and livestock to be processed should be properly managed to avoid polluting foods. Their feeding area should keep a certain distance from the workshop and should not be located within the prevailing wind direction. The factory area should have a smooth drainage system without any possibility of serious waterlogging, leakage, silt, dirt, breakage, or breeding of harmful animals that could cause food contamination. The surroundings of the factory should be adequately designed and constructed to prevent the entry of pollutants. If a fence is required, airtight construction materials should be used in the section at least 30cm below the ground. If the restaurant, staff quarters, and other living areas are attached to the factory, these should be separated from the workplace and storage places for food or raw materials.

4.3.2 Factory Buildings and Equipment

New construction, expansion, and alteration projects (such as food factories and workshops) should be undertaken according to the established standards. Factory settings should include the production workplace and auxiliary area. Factory settings should meet the needs of the production process and hygiene requirements: orderly, neat, scientific layout, rational convergence of processing steps, and preventive measures against cross-contamination between raw materials and semifinished products, raw materials, and delicatessen food. The configuration and usable area of the workshop and storage sites should be adapted to the quality requirements, as well as the variety and quantity of products. Per capita area in the workshop should not be $<1.50\,m^2$ (not including the equipment-occupying area), and the height should not be $<3\,m$. There should be appropriate channels or workspaces (the width of the general workspace should be $>90\,cm$) between the equipment and between the walls and equipment in the workshop to ensure that the operation staff (including those for cleaning, disinfection, and mechanical maintenance) will not contaminate the food or packaging materials because of clothing or body contact. It is necessary to establish independent physical, chemical, and microbiological laboratories with sufficient space. When necessary, establish independent visual inspection rooms and sample rooms that are equipped with corresponding testing equipment.

4.3.3 Workshop Isolation

The workshop should be isolated in accordance with the production process, production operation requirements, and cleanliness requirements in the operation area to prevent cross-contamination. The areas with different levels of cleanliness (such as clean, quasi-clean, and general work area) should be isolated (Table 4.5). According to the natural sedimentation method in the GB/T 18204.1-2000, the total number of colonies in the air of the production work area should be controlled; the number of colonies per unit (CFU/unit) in the general work area should be ≤500 CFU/unit, that in the quasi-clean work area should be ≤75 CFU/unit, and that in the clean work area should be ≤50 CFU/unit.

4.3.4 Structure of the Factory Buildings

The factory buildings should be built with appropriate building materials that are strong and durable and easy to repair and clean and can prevent the contact surface of raw materials, semifinished products and foods, and inner packing materials from contamination (such as pest invasion, inhabitation, and reproduction).

To prevent cross-contamination, personnel and material transport channels should be set up separately and equipped with air curtains (i.e., wind curtain) or water curtains, plastic curtains, or bidirectional spring doors. There should be a buffer chamber for each staff and material transport channel between the areas with different cleanliness levels. If there are stairs or tracks crossing the production

Table 4.5 Cleanliness levels in each workplace for food companies

Factory setting	Cleanliness level	Colony requirements in air (CFU)
Raw material warehouse	General work area	≤500
Material warehouse		
Raw material-processing sites		
Empty bottles (cans) stacking and sorting places		
Inner packaging container cleaning places[a]		
Sterilization places (using sealed equipment and pipeline transportation)		
Closed fermentation tank		
Outer packing chamber		
Final product warehouse		
On-site laboratory		
Other relevant auxiliary areas		
Manufacturing site	Quasi-clean work area	≤75
Inner packing chamber for nonperishable instant foods		
Inner packing preparation room		
Buffer areas		
Other auxiliary areas		
Cooling and nonenclosed storage of perishable instant foods and final semifinished products	Clean work area	≤50
Inner packing chamber of perishable and instant food products		
Microbial inoculation room		
Other auxiliary areas		

Note: Professional norms prevail.
[a] *The outlet of the inner packaging container cleaning places should be set up within the restricted operation zone.*

line in the workshop, these should be prevented from polluting nearby foods and contact surfaces of foods. The construction of safety facilities is also necessary.

4.3.5 Safety Facilities

Ground wire and leakage protection socket systems must be installed to supply power to the factory. Different voltage outlets must be clearly marked. The sockets used in a high humidity environment should have waterproof function. Fire and explosion protection facilities should meet fire safety requirements. When necessary, emergency equipment should be set up in proper and obvious places.

4.3.6 Roof and Ceiling

The materials used in the interior roof or ceiling of the workshop and warehouse should have a smooth surface and be nonabsorbent, odor free, easy to clean, corrosion resistant, heat resistant, and light colored. There should be a proper roof slope or curvature to reduce and prevent condensation dripping. Installation of air conditioning ducts in the workshop should comply with food hygiene requirements, and they should be installed above the ceiling.

4.3.7 Walls, Doors, and Windows

The materials used for the walls of the restricted operation zone should be nontoxic, odor free, smooth, impervious, easy to clean, anticorrosive, and light colored. The use of lead paint should be avoided. White ceramic tiles or other anticorrosive materials can be used to decorate the dado at a height of no higher than 1.5m. In the restricted operation zone and humid environment, connections between the walls, the walls and ceiling, and the walls and ground should have proper radii (curvature radius should be >3cm) to facilitate cleaning and disinfection.

All doors and windows in the workshop should be installed with sealing frames that are rustproof and waterproof and easy to clean instead of the wooden doors and windows. It is essential that all windows could be opened during the working process and installed with a rustproof and double-layer gauze of >26 mesh that can be easily disassembled and cleaned. Windowsills should not be built in the workshop. If there are windowsills, these should be located 1m above the ground, and the mesa should tilt inward by approximately 45°.

At the entry and exit points of the restricted operation zone, separate buffer chamber and facilities (such as air curtains, plastic curtains, and automatic closing doors) and/or sole cleaning and disinfecting equipment (shoe-changing facilities should be set in the workplace that should be keep dry) should be installed. Doors should be made of smooth, solid, and waterproof materials and should be kept closed. The insulating materials in the workshop should have no asbestos-containing materials.

4.3.8 Ground and Drainage

The ground should be paved with building materials that are nontoxic, nonabsorbent, antiskid, and corrosion resistant. They should prevent water seepage, have no cracks, and be easy to clean and disinfect (such as acid-proof brick, terrazzo, and concrete). The appropriate grade of slope is 1–1.5%. In the area where there is liquid flowing to the ground, the environment is often wet or needs to be cleaned using water; the slope grade should be designed within the range of 1.5–3% according to the size of the flow. Sinkholes should be constructed on the ground but not directly below the production equipment. All drainages should be provided with water storage elbows and a filter mesh of appropriate size to prevent odor and solid wastes from clogging the drainage.

The junction of the side and the bottom of the drains should have appropriate radii (curvature radius should be >3 cm). The gradient should be approximately 3%. The direction of the flow should be from a high cleaning area to a low cleaning area, and a design to prevent backflow should be used. In addition, a device in the drain outlet to prevent the invasion of harmful animals should be installed. Wastewater should be discharged to the wastewater treatment system or disposed using other appropriate means.

4.3.9 Water and Steam Supply Facilities

The production water (ice) must follow the provisions of GB 5749-2006. Water quality, pressure, flow, and other indicators must meet the needs of normal production. When necessary, water storage equipment in the workplace and hot water of appropriate temperature should be provided. The water supply system should have measures to prevent backflow and allow siphoning. Moreover, there should be a diagram of the complete water supply network.

The water storage device (pool, tower, or trough) and the water supply pipeline that are in direct contact with the water appliances should use nontoxic, odor-free, and anticorrosive materials. Safe and hygienic sanitation sites should be constructed at the entrances of water supply facilities to prevent the entry of harmful organisms and substances. If self-provided water instead of tap water is used, appropriate water purification or disinfection facilities should be constructed (such as sedimentation, filtration, removal of iron, manganese, and fluoride, and sterilization), according to the characteristics of local water to ensure that the water quality meets the requirements of GB 5749-2006 and other related standards.

The piping system of nonpotable water that is not in direct contact with foods (such as cooling water, sewage, or waste water) should be clearly distinguishable using different colors and used as completely separated pipes. Backflowing or mutual transferring of this water is forbidden. Steam in food production, regardless of its use, must comply with food hygiene requirements. Steam that is directly injected or in contact with food should be filtered, and any debris should be removed.

4.3.10 Lighting and Ventilation Facilities

Lighting facilities should be installed according to the provisions of GB 14881-1994 4.5.9.

Processing, packaging, and storage places should have excellent ventilation to prevent excessively high temperatures, steam condensation, or odors within the facilities. When necessary, these places should be equipped with ventilation equipment. The direction of airflow should be from the high cleaning area to the low cleaning area. Air access to the restricted work zone should always be clean. A stable temperature and humidity should be maintained in the clean work area.

Ventilation devices should be removable for cleaning, repair, or replacement. Vents should be equipped with anticorrosive net covers. Air inlets must be 2 m away from the ground and far away from potential sources of pollution. The exhaust ports should be well equipped and have devices that prevent the invasion of harmful organisms. There should be excluding, collecting, or controlling devices in the workplace that could be used during incidents of odor, gas (steam and poisonous and harmful gas), dust, and food contaminations.

4.3.11 Hand-Washing Facilities and Disinfection Pool

Hand-washing facilities should be made of stainless steel or ceramics such as impervious materials that are easy to clean and disinfect. These facilities should be provided at the entrance of the workshop and at various appropriate locations using nonmanual faucets (including pressing and automatically closing and elbow-operated types). When necessary, warm water of appropriate temperature should be provided (or hot and cold water and faucets that can be adjusted to provide hot and cold water). The number of faucets should meet the needs of workers. There should be liquid disinfectants at the hand-washing basins and straightforward instructions on hand washing. The hand-washing facilities should include a dispenser for detergent and disinfectant, as well as a hand dryer or hand towel. The chlorine-containing disinfectant used for hand disinfection should have a free chlorine concentration of 50 mg/kg. Water from the hand-washing facilities should flow directly into the sewer and be equipped with a device to prevent backflow, harmful animal intrusion, and odor. The entrance of the restricted operation zone should set up a footwear disinfection pool or sole-cleaning facility. (For the footwear disinfection pool, the concentration of free chlorine in the disinfectant should be 200 mg/kg.) The workplace should be kept dry and clean, with provisions for a shoe-changing area. The shoe disinfection pool should be designed according to the provisions of GB 14881-1994 4.6.1.5.

4.3.12 Locker Room

Locker rooms should be located at the entrance of the workshop and close to the hand-washing facilities. Locker rooms for men and women should be separate. The size of each locker room should conform to the number of production personnel. Lighting and ventilation in locker rooms should be in a good condition, and these rooms should be equipped with disinfection devices. The locker rooms should be divided into external and internal rooms. The locker room should have sufficient closets, shoe racks, and mirrors for individual use by the production staff. Wardrobes should be numbered, and the slope of the top of the cabinet should be >45°. The cabinet should be equipped with clothes hangers that are not leaning against the wall. The locker rooms should have independent shower rooms that are designed according to the provisions of GB 14881-1994 4.6.3.

4.3.13 Toilet

Toilets should cater to the needs of the production staff and must always be in a hygienic state. The number of toilets and urinal pits should correspond to the size of the production staff and personnel. Large channel flushing toilets are prohibited. The pit latrines in the factory should be >25 m away from the production workshop and should always be clean, with mosquito and fly prevention facilities set up in the area. The toilets located in the production area should be outside the workshop and connected to the locker rooms. The entrance of the toilets should not face the entrance of the workshop. The toilet door should not open outward, and an automatic closing device should be installed. The toilet should have hand-washing facilities, as well as excellent ventilation and lighting. The ground, walls, ceilings, partitions, and doors should be made of easy-to-clean and impervious materials. The toilet sewage and drainage pipes should be separated from those in the workshop and have sturdy odor and gas seals.

4.3.14 Warehouse

The warehouse (cold storage) should be built to meet the production capacity and the requirements of product storage. Its size should ensure the factory's smooth operation and should be easy to clean. The warehouses for raw materials and final products should be separate. Different items in the same warehouse should be appropriately segregated. The warehouse should be built from nontoxic and durable materials and have measures to protect the stored goods from pollution. The warehouse must have devices that prevent the invasion of harmful animals (such as the anti-rat boards or anti-rat ditches at the doors). The warehouse should have a sufficient number of stacks or shelves. An appropriate distance should be maintained between the storage objects and walls and the ground to maintain air circulation and transport of goods. The warehouse should have temperature and humidity recorders for different storage requirements of the facility. There should be a special storage area for dangerous materials such as insecticides, acids, alkaline solutions, and other toxic and harmful materials. The storage area for dangerous materials should be far away from the production workshop and food warehouse. Cold storage areas should address the different humidity and temperature requirements of the materials that need to be stored.

4.3.15 Mechanical Equipment

1. Design

All food-processing equipment, including pipes and instruments, should be easy to clean and disinfect, as well as simple to monitor. The equipment design and structures should not be exposed to machine-lubricating oil, metal debris, sewage, and other pollutants in order to prevent their transmission to the food. The contact surfaces of equipment used in processing foods should be smooth or corner smooth, with no blind angles and cracks to prevent the accumulation of food debris, dirt, and organic matter. The design and manufacture of the storage or transport systems should be easy to maintain and disinfect. In the food production or processing area, all equipment and utensils that are not in contact with foods should always be kept clean.

2. Texture of materials

Food production and equipment, operation platform, conveyor belts, transport vehicles, engineering equipment, and other auxiliary facilities that may be in contact with food should be made of nontoxic, odorless, nonabsorbent, corrosion-resistant, and easy-to-clean and -disinfect materials and should comply with the national mandatory standards. Surfaces in direct contact with foods should not be made of materials that may bring potential harm or contamination such as bamboo or wooden materials. Lubricants used in the equipment that come in contact with food must be food grade. All outer layers of the insulation facilities in the plants must be made of nonabsorbent materials.

3. Setup and installation

The setup should comply with the provisions of GB 14881-1994 4.4.3.1. Installation of equipment should comply with the provisions of GB 14881-1994 4.4.4.1.

4. Production equipment

Enterprises should have production equipment adapted to its products and the processing technology. The processing capacity of different equipment should be mutually supporting. The production equipment layout should meet the process requirements and ensure smooth production in an orderly manner to avoid cross-contamination. Measuring and recording instruments used to measure, control, or record should be able to generate accurate data and perform periodic calculation. The enterprise should have adequate air supply equipment to ensure proper ventilation during drying, conveying, cooling, and purging. The contact surface of food and the compressed air of the food surface should be cleaned to remove oil, moisture, dust, bacteria, insects, and other debris.

4.3.16 Quality Inspection Equipment

All inspection equipment should meet the standards of daily quality and hygiene inspection of raw materials, semifinished products, and final products. The test apparatus and equipment must be inspected on a regular basis and undergo regular calibration to ensure accurate test results. When necessary, indices that could not be detected using the factory's methods can request an inspection agency to perform testing.

4.3.17 Management Mechanism and Personnel

Organization and responsibility are shown as follows:

A quality management agency that would lead the company (plant) or the supreme leadership of enterprises directly under the group should be established. This quality management agency will take overall responsibility for enterprise quality management. The enterprise should set up production, quality, hygiene, and other functional management departments. No single person will be responsible for two management positions. The production management department is responsible for raw material processing, production and product packaging, and other operations related to production. The quality management department is responsible for quality control standard formulation, sampling inspection or quality tracking of raw materials, packaging materials, and processing and production, and other relevant quality-related activities. The hygiene management department is responsible for the formulation and revision of hygienic schemes in the inner and outer environment and plant facilities; management of hygiene in production, cleaning operations, and of personnel; and hygiene training and health examination of employees. The quality management department should have full authority of quality management implementation, and the responsible person should have the power to stop production and delivery of final products. This department should have food inspection staff members who are responsible for quality inspection and analysis of raw materials, semifinished products, and final products. A hygiene management team consisting of responsible persons from the hygiene management, production management, and quality management departments should also be established. They are responsible for hygiene planning, audit, supervision, and assessment. This team should be equipped with full- or part-time personnel with professional training who will be responsible for the management of food hygiene, the propagation and implementation of food hygiene regulations and other relevant rules, and the supervision of the hygiene system and relevant records.

4.3.18 Personnel Requirements

The personnel in charge of the enterprise should be well-versed with the "Health Law of the People's Republic of China" and "Product Quality Law of the People's Republic of China" and other relevant laws and regulations and should have professional knowledge of food safety and health, production, and processing. The responsible persons of the production, quality, and health management departments should also be familiar with these laws and regulations and should have college-level professional degrees of related majors, mid-level or higher titles, or secondary professional degrees of related majors and direct or related experience in management higher than 4 years. The responsible persons for production management should have corresponding knowledge of techniques, production technology, and hygiene. Quality management personnel should have the ability to identify adverse situations in each production line. Food inspection personnel should have a college diploma or higher, must have graduated from a secondary technical school and have been engaged in food inspection work for at least 2 years, or possess relevant qualification in professional inspection after his or her professional training has been approved by the administrative departments at the provincial level or higher.

The enterprises should have a sufficient number of personnel to meet the requirements of quality management and product inspection throughout the production process. Professional health management personnel should possess a professional college degree or equivalent health-related educational background. Part-time health management personnel should have a secondary specialized technical school degree or higher, or equivalent health-related educational background.

4.3.19 Education and Training

The enterprise should have a plan to provide various department managers and employees different preservice and in-service training to increase their knowledge and skills.

4.3.20 Health Management

1. Management requirements

The departments in an enterprise should formulate an appropriate health management system according to the standard contents audited, supervised, and implemented by the leadership group of the health management department. The health management department is in charge of the establishment and implementation of the inspection plan. The team of health management personnel must check the health system status of the facilities on a daily basis. The health management personnel must check the hygiene of production and environment within the factory at least once a month. Each inspection should be recorded and documented.

2. Environment hygiene management

The plant and adjacent areas should be kept clean. Roads and grounds should show no damages, waterlogging, or dust. Trees must be pruned regularly to keep the environment clean and tidy. Piled debris and unwanted equipment are prohibited to prevent the breeding of harmful animals. The drainage system should be smooth and show no sludge sedimentation, and the waste should be properly disposed. Harmful (toxic) waste gas, wastewater, noise, and other harmful effects on the environment should be avoided. Temporary waste storage facilities at suitable locations separate from the food production workshop should be constructed. Waste should be classified and stored according to their properties. The storage facilities for perishable wastes should be closed or covered. Dirt should not be splashed and must be discarded on a daily basis. All wastes should be cleaned and discarded from the factory at least once a day. Containers should be cleaned and disinfected upon removal of its contents. Waste space should not emit a bad odor or harmful (toxic) gas. Breeding of harmful animals should also be prevented. Contamination of food, food-contacting surfaces, water sources, and the ground should also be prevented.

3. Factory facilities and hygiene management

A facility repair and maintenance system should be established in accordance with the provisions to ensure that the building is in a good hygienic condition. The facilities in the workshop should be kept clean, functional, and up-to-date. When the roof, ceiling, and walls are damaged, immediate repair is highly required. The ground should have no damaged areas and water logging. The pretreatment sites of raw materials, processing and manufacturing sites, toilets, locker rooms, shower rooms, and other locations (including ground, ditches, and walls) should be cleaned and disinfected every day before and after work. When necessary, increased cleaning and sterilizing may be performed. Hand dryers should be regularly controlled and inspected to avoid becoming the source of pollution. A specific responsible person should manage the toilets in the workshop. The appearance of lamps and their piping system should be cleaned regularly. Cold storage should be always cleaned and regularly disinfected to avoid water logging on the ground and the emergence of mildew on the walls. The temperature in the cold storage should be regularly monitored and recorded using automatic temperature-recording instruments. Workplaces should have devices (such as window screens, curtains, fences, and trap lamps) to prevent the invasion of harmful animals. Pest control in plants should be executed in accordance with relevant provisions. In the appropriate positions of workplaces for the handling of raw materials, processing, packaging, and storage, the containers that are impervious, easy to clean and disinfect (except the disposable ones), and capped (or sealed) are installed for waste storage and should be regularly (at least once a day) moved out of the workplace. Commonly used containers should be regularly cleaned and disinfected after the contents have been discarded. Waste treatment equipment should be regularly cleaned and disinfected during downtime. Raw materials, packaging materials, or other items should be used immediately after receipt. The restricted operation area should not be stacked with rarely used goods, packaging materials, or other unwanted items. The production workshop should not store toxic substances. Cleaning and disinfecting supplies used inside the workshop should be stored in specific areas or cabinets and should be clearly marked and managed by specific persons.

Water storage tanks (tower and pool) in the workshop should be cleaned regularly, and their disinfection state should be checked every day before starting work. If using self-provided water, the water should be tested by an inspection agency twice a year to ensure that the water quality conforms to the standards of GB 5749-2006. If the total number of colonies in the air does not meet the requirements of the corresponding clean area, then the air should be disinfected. The frequency of inspection of microbial colonies in the air in the production area is 1–2 times per week for the clean work area and 1–2 times per month for the quasi-clean work area. When the test results vary, the frequency of inspection should be increased, e.g., 1–2 times per season for the clean work area. In addition, in spring and summer/rainy seasons, more frequent inspections are necessary.

4. Machinery and equipment hygiene management

Equipment, tools, and pipes for manufacturing, packaging, storage, and transportation should be regularly cleaned and disinfected according to relevant provisions for cleaning and disinfection. All food-contacting surfaces, including those of equipment and utensils, should be regularly disinfected and thoroughly cleaned after sterilization (except for thermal disinfection) to prevent disinfectant residues from contaminating the food. Used equipment and tools should be thoroughly cleaned and disinfected. Cleaning and washing are necessary before the next food-processing cycle (except for equipment and tools that come in contact with dry foods). After using pesticides in the workshop, all equipment and utensils should be thoroughly cleaned to remove residual drug and pollutants. The mobile equipment and utensils that have been cleaned and disinfected should be placed in sites where the food-contacting surface will no longer be contaminated and be always ready to use. Clean water used for cleaning equipment and utensils that are in contact with foods should meet the requirements of GB 5749-2006. The filtration system of compressed air should be regularly maintained to prevent contamination and to ensure the hygiene and the quality of compressed air. Mechanical equipment and sites for food production should be used exclusively for this activity.

5. Auxiliary facility hygiene management

(1) Water supply station within the enterprise

Detailed rules of operation and management should be formulated with the strict systematic water quality inspection, system repairing, and maintenance of records. Supervisors should regularly evaluate these rules (at least once quarterly). The utensils must comply with hygiene requirements. All equipment should be regularly maintained and kept clean. Disinfectants must be properly stored with strict registration procedures and recorded. Other independent sundries that are not directly related to water treatment should not be placed in the station. Water storage tanks (tower and pool) should be regularly (at least once quarterly) cleaned and disinfected, and the water quality should be checked at any time to ensure that it conforms to the provisions of GB 5749-2006. The water treatment equipment should be regularly or frequently cleaned and maintained according to the actual operation conditions. Unauthorized personnel should not enter the water supply station. All access ports, doors, and windows must remain closed.

(2) Boiler room

Boiler room operating personnel should undergo the necessary training and certification for the safe operation, maintenance, and repair of the boiler in accordance with the relevant management requirements. The chemicals for water treatment of the boiler must be nontoxic, the dosage must be strictly controlled, and the sewage must be periodically emptied (with corresponding record-keeping). The exhaust of the boiler should be monitored to ensure that its emissions comply with the GB 13271-2001 requirements. Exhaust pipes should be regularly cleaned to prevent pollution in the factory environment.

6. Cleaning and disinfection management

The enterprise should formulate a system of cleaning and disinfecting to ensure all places, equipment, and utensils are kept clean. All agents used in cleaning the food equipment, utensils, and packaging materials must be food grade. Nonfood-grade cleaning agents that may endanger food safety and hygiene are prohibited. The method for cleaning and disinfecting must be safe and sanitary.

Disinfectants and detergents must be in a secure and usable state. Cleaning and disinfection equipment and instruments should be placed in the special safekeeping place and managed by a specific responsible person. Employee health management should implement the regulations of GB 14881-1994 5.12.

7. Personal hygiene management

Operators of food production must maintain good personal hygiene and should regularly wash his or her hair, cut nails, shower, and change clothes. Before entering the production workshop, they must wear clean work clothes, hats, shoes, or boots. Overalls should be used to cover the coat. Hair should not be exposed, and a mask should be worn when necessary. Personnel should not wear their work clothes and shoes in the toilet or outside the workshop. During operation, hands should be clean. Hand washing and disinfection should be performed before working at all times. Hands should be washed and disinfected during the following scenarios: before starting work, after going to the toilet, after the treatment of contaminated raw materials and items, and after engaging in other activities unrelated to production. The enterprise should include supervisory measures on personal hygiene. The staff in direct contact with foods should not use fingernail oil, lipstick, powder, and other cosmetics and should not wear watches, rings, necklaces, earrings, and other accessories. Before starting work, they should not drink alcohol. Smoking in workplaces is strictly prohibited. Unrelated behavior such as eating and doing something that influences the maintenance of food hygiene is also prohibited.

Operators with hand injuries should not come into contact with foods or raw materials unless a bandage covers the wound and protective gloves are worn. Personal clothing, work clothes, and other nonwork-related items should be stored elsewhere in the locker room and are not allowed into the workshop. Visitors' access to workplaces should be managed properly. Whoever wants to enter the restricted operation zone should follow the personnel hygiene requirements of the field workers.

8. Pest eradication management

A pest eradication management scheme should be implemented at the enterprise. Effective measures that prevent the entry and breeding of rodents, mosquitoes, flies, and insects should be taken in the factory. Potential places where harmful animals could live should be controlled and eliminated. The enterprise should set up a mouse-killing map and be equipped with the necessary mouse-killing facilities and a specific person who is in charge of checking and recording every day. Sticky paper, cages, and traps can be used for killing mice; however, the application of toxic mouse-killing chemicals or agents must be prohibited. If harmful animals are found, the source should first be identified to eliminate potential hazards. Killing harmful animals must ensure no contamination of foods, food-contacting surfaces, and packaging materials (such as avoiding the application of pesticides). Only when other control measures are invalid can pesticides be used. Before using pesticides or other types of poisonous agents, measures for preventing pollution and poisoning of personnel, foods, equipment, and instruments should be taken. After using pesticides, all equipment and utensils should be thoroughly cleaned to eliminate contamination. Pest-eradicating work cannot be performed during the production process. All raw materials and final products must be protected from contamination with pesticides. Pest-killing lamps may be set up at the entrance of the workshop and within the workshop to kill insects and flies that may enter the workshop. The pest-killing lamp in the workshop must be located far away from the production area. Pest-killing lamps should be cleaned every day.

9. Sewage management

Sewage discharge should comply with the requirements and standards of GB 8978-1996. The enterprise should have a detailed sewage-discharging network. The daily sewage treatment capacity must match the actual production scale.

10. Working clothes management

The design and materials of working clothes must meet the requirements for food hygiene. Working clothes should include light-colored jackets, pants, hats, shoes, and boots. Masks, aprons, sleeves, and other hygiene protection products should also be used during food processing and production. Workers who are in direct contact with the products must change their clothes daily, and other workers who are not in direct contact with the products should also change their clothes regularly. Staff engaged in the coarse processing and fine processing area, restricted operation zone, and other operation areas must wear clothes and caps of different colors in order to distinguish clothes used for each area. There should also be a washing and cleaning system for clothes. Working clothes must be cleaned and disinfected in the washing room and managed by a specific responsible person. Cleaning of working clothes must be separated according to the restricted operation zone and other operation areas so as to prevent cross-contamination.

11. Management of toxic and hazardous substances

Management of toxic and hazardous substances should be executed according to the regulations of GB 14881-1994 5.6.

12. Production process management

Production process management includes the formulation and execution of production and operation rules, which should include the following:

- Product formula.
- Standard operating procedure.
- Regulations of production management (including at least the operation process, management objects, monitoring projects, monitoring limit, and monitoring standards and attention).
- Procurement standards of raw materials.
- Operation and maintenance standards of machinery and equipment.

1) Raw materials and packaging materials

Raw materials and packaging materials should conform to the established standards. Raw materials, production environment and process, and quality control of semifinished products (used as raw materials) inside or outside the factory should follow the established GMPs. Fresh fruits, vegetables, and other agricultural products that easily spoil should be purchased on the basis of production needs and purchasing plan and be processed as soon as possible after entering the factory from the original or supply sites. Untreated raw materials should be refrigerated or stored in the shade and in a properly ventilated place and should be checked using sensory detection methods before use. Foreign materials that do not meet the quality and hygiene requirements should be discarded. The qualified and unqualified raw materials should be stored separately, with a clear and straight sign. Raw materials should be stored in accordance with the production rules or relevant standards to avoid contamination and damage. Damaged outer packaging materials should be stored separately, labeled with the specific type of damage, and used only after passing quality inspection. Frequently used materials (such as returned materials) or those continuously used should be stored in clean and sealed containers and clearly labeled. Stagnant water should not be used for cleaning raw materials. Washing water should not be recycled to avoid secondary pollution. Frozen raw materials should be thawed to prevent contamination and deterioration of quality. Stagnant water should not be used in thawing frozen raw materials. When production has been completed, unused materials should be properly stored in the appropriate storage place to prevent pollution and must be used in the next processing cycle, taking note of its expiration date.

2) Production process

Production operations should be in compliance with the principles of safety and health and should be executed to reduce harmful microbial growth rate and food pollution. The physicochemical conditions (such as time, temperature, water activity, pH, pressure, and flow rate) and processing conditions (such as freezing, refrigerating, dehydration, heat treatment, and acidification) in food processing should be strictly controlled to ensure that in the event of a mechanical failure, time delay, or temperature change, food spoilage or pollution will not occur. Perishable foods should be stored in accordance with the production rules or established standards. Effective measures should be taken to prevent the exposure of foods to secondary pollution in the production process or during storage. Equipment, containers, and tools used for conveying, loading, and storing raw materials, semifinished products, and final products, as well as their operation, usage, and maintenance should be prevented from polluting foods during processing or storage. Equipment, containers, and utensils that had been in contact with raw materials or pollutants must be thoroughly cleaned and disinfected; otherwise, they should not be used for food processing. All containers for holding semifinished products should not be placed directly on the ground or on a wet surface that had been contaminated to prevent contamination caused by water splashes or indirect pollution from outside contamination from the bottom of the container. The circulating cooling water in direct or indirect contact with product packaging materials should be kept clean and replaced regularly. The ice cubes in direct contact with foods should be produced under sanitary conditions. Effective measures (such as screening mesh, traps, magnets, and electronic metal detector) should be taken to prevent metal debris or extraneous materials mixing with foods. The processing temperature and pretreatment heating time of foods should be strictly controlled and cooled down within the specified time and quickly moved to the next process. Regular cleaning for the heating pretreatment equipment is necessary to prevent the growth of heat-resistant bacteria. The activity of water should be controlled to prevent the growth of harmful microorganisms in foods, and the water should be treated. The pH should be controlled to prevent the growth of harmful microorganisms in foods, and the pH should be adjusted and maintained below 4.6. Packaging materials must be cleaned and sterilized before use. During transportation and sales, inner packaging materials should effectively protect the foods from harmful substances and follow the established standards of food hygiene. Flushing in a large area during the production process should be avoided. When necessary, the nozzle must be lowered as much as possible and close flushing must be conducted to reduce water splashes. Welding, cutting, and grinding should not be executed in the production process to prevent the production of peculiar smells and debris. Pest-eradicating measures should not be executed in the production process. The windows in the clean work area should be closed during production. Daily equipment maintenance should be strengthened to keep the equipment clean and hygienic. The correct operating procedures must be strictly enforced in the maintenance of equipment. Equipment failure should be immediately addressed to prevent affecting the quality and hygiene of the final products. The equipment should be checked to determine whether it is in a normal functional state before each production process. All production equipment should be regularly inspected, and the repair records should be maintained.

4.3.21 Quality Management

The enterprise should completely collect the standard documents on food quality, hygiene, and other relevant documents regarding the established laws and regulations.

1. Quality management of raw materials and packaging materials

The detailed procurement system, including supplier assessment, quality specifications, test items, acceptance criteria, and inspection methods should be formulated. The operation procedures of weighing, sampling, inspecting, judging, auditing, and processing should be established and strictly implemented. The suppliers should provide inspection and quarantine certificates or laboratory test sheets during the purchase of all raw materials. Each batch of raw materials should not be used prior to quality verification. Rejected raw materials should be marked (unqualified or banned), stored, and immediately treated. The qualified raw materials should be used in accordance with the principle of "first in, first out." The raw materials should be numbered by batch according to the production date and the number of suppliers. The batch number should be used in the production records to facilitate tracing. Opening of packaging containers of raw materials should be immediately performed to prevent deterioration of package contents. Raw materials with a longer storage period and potential quality change should be sampled for quality verification prior to use. Raw materials that do not meet the requirements should not be included in the production process. Storage of raw materials that require special storage conditions should be controlled and recorded. Raw materials may contain pesticide residues, drug residues, heavy metals, or mold and toxin, and the contents of these substances should be confirmed within the national standards. Raw materials with special requirements such as caffeine should be used and stored according to relevant provisions. Food additives should be stored in special warehouses or cabinets and managed with strict material-using methods and within the effective period by a specific person. A special registration book should be used to record the types, purchase volume, and usage. The application of food additives should comply with the provisions of GB 2760-2011. Nitrates and nitrites should be kept separate from the general food additives and clearly labeled.

2. Process quality management

The enterprise should adopt HACCP management and strictly enforce the production procedures. Formulation and process conditions should not be arbitrarily changed without approval. Quality issues in the production process should be promptly tracked and corrected. To monitor the quality of each step during the production process and to facilitate tracing, the enterprises should sample semifinished products at the control points of the production process and prepare quality records, production records, and other management reports. Unqualified semifinished products are not allowed to proceed to the next process and should be properly treated with corresponding treatment records. Regular inspection of work tables, tools, work clothes, and operators' hands should be executed, and microbial sampling tests (aerobic bacterial count and coliform bacteria) of the final products after packaging and other products should be conducted. The presence of mildew, yeast, and fungi should be checked regularly. Despite knowledge of correct and completed cleaning and disinfection, the work should still be validated. Under normal circumstances, regular checking should be executed once a week. If it is not required, checking should still be performed every day until it is qualified. It must be verified again before starting production after the shutdown. Inspection records for each batch of final products before storage should be complete. Any unqualified items should be properly treated and documented.

3. Quality management of final products

Detailed product quality indexes, inspection items, inspection standards, and sampling and testing methods should be formulated. The lowest limit of quality indexes may not be lower than the national standards. Testing methods should be based on the national standard methods. A non-state standard testing method should be regularly checked with the national standard method. The final products should be randomly sampled batch by batch, and delivery inspection should be conducted according to the product standard. Unqualified products should not be delivered, and treatment records should be prepared.

The analysis results should be filled in the "product quality inspection record sheet" with "production records" to determine whether the final products are qualified for the approval of delivery at the same time. Management and records of storage conditions should be reviewed after the final products enter the warehouse. When the final products are delivered, the production date, shelf-life, and external quality should be checked again. Vehicles that cannot maintain the good quality of the products must not be used for transportation. The sampling plan of final products should also be formulated. Samples from each batch of final products should be inspected in terms of quality when necessary. When necessary, the shelf-life of the final products should be monitored.

4. Storage and transportation management of final products

Storage and transportation of goods should avoid direct sunlight, rain, severe temperature and humidity, and collision to maintain product quality. The final products should be stored separately according to their varieties, packaging styles, and production dates, following the FIFO (first in, first out) principle. The warehouse should be well organized. Goods should not be placed directly on the ground. The final product warehouse should not store toxic, harmful, perishable, and inflammable items, as well as items that may cause odor. Goods in the warehouse should be checked regularly. Abnormal states should be handled immediately. The temperature records in the warehouse should be maintained (and humidity records when necessary). Products with damaged packages or potentially large deteriorations in quality due to extended storage should be reexamined. Various transportation tools and vehicles of raw materials and final products should be kept clean and disinfected regularly. When transported, toxic, harmful, corrosive, or odorous goods should not be mixed. Non-van transportation tools and vehicles should be equipped with dust-proof, anti-sunlight and rain canvas, plastic cloth, and other coverings. Goods with special requirements should be transported using specific vehicles such as insulation trucks and refrigerated trucks. The storage and transportation of raw materials and final products should be recorded in detail, including batch number, delivery time, place, object, and quantity to assist in any recalls when quality issues emerge.

5. Customer service management of final products

Enterprises should establish a system for handling complaints from consumers. The responsible person in the quality management department should immediately investigate the reasons behind the consumers' complaints (when necessary, they should coordinate with other departments) and execute improvement plans or strategies. The enterprise should send the relevant personnel to explain the situation to the consumers or to apologize to them and put forward the treatment and rescue measures at the same time. Enterprises should establish a product recall system. Consumers' complaints and the recalls of final products should be clearly recorded with the product name, batch number, quantity, causes, treatment date, and final treatment results. The statistics and analysis of the records should be conducted regularly and sent to the relevant departments to improve the future work.

4.3.22 Record Management

Besides regularly recording and checking the inspection results, the responsible person in the hygiene department should also complete the tables of daily health management records, including the implementation of cleaning and disinfection, personnel health status, treatment results of abnormal situation, and prevention measures to avoid the recurrence of these abnormal situations. Quality management should have a detailed record regarding the quality management activities and results of the whole process from the input of raw materials to the delivery of final products, which will help in checking and comparing with the originally planned goals.

The production department should report production and management records and have a detailed record of responses to abnormal situations and preventive measures against the recurrence of these abnormal situations.

These records should be reviewed and signed by the executive personnel or relevant management personnel. Records must be true and synchronous with on-site inspection or monitoring, not in advance or afterward. The records must be standardized and clear. If the record is changed, the original record should not be altered and the person who amended the document must sign this document.

The health, production, and quality management departments should separately audit records of health, production, and quality management, respectively. If there is an abnormal situation, the immediate corrective action should be taken. The preservation period of relevant records prescribed in this standard should be extended by 6 months beyond the expiration date of the product, at least 2 years.

4.3.23 Tag

Product labels and instructions should meet the requirements of GB 7718-2011.

Retail products should be labeled with the following information in Chinese and using general symbols:

Product name and product standard number; ingredients including names, contents, and food additives; net content: weight, volume, or quantity; enterprise name, address, telephone number for consumer service, telephone number of production enterprise; and shelf-life, which should be printed instead of being marked with labeling; batch number, the code or password to reveal the production batch number, and the original data of the products can be traced according to the production batch number; introduction of method of consumption and storage conditions; other indicators prescribed by relevant departments; the relevant batch number should also be marked on outer packaging containers, which is convenient for storage management and product recall; and the packaging and shipping mark should conform to GB/T 191-2008.

4.3.24 Establishment and Evaluation of the Management System

Enterprises should establish a comprehensive and effective management system for the implementation of this standard, which will coordinate each department of the enterprise to implement the rules and regulations of this standard.

The enterprise should establish an internal assessment group that consists of all levels of management to execute regular or irregular inspections on the implementation of this standard and to resolve and track the problems in a reasonable manner. The members of the internal assessment group should be subjected to training, and training records should be prepared. The enterprise should formulate internal assessment plans with designed inspection and assessment periods (generally conducted every 6 months), and implementation records should be completed. Enterprises should establish operating procedures for the formulation, revision, and abolishment of relevant management regulations in this standard to ensure that the quality management personnel has a valid version of the management files and execute according to the effective version.

4.4 CONSTRUCTION OF THE TRACEABILITY MANAGEMENT SYSTEM FOR CANNED CITRUS PRODUCTS

Traceability refers to the ability to trace the source, function, location, and history of a product using established records.

The terms involved in system establishment include distributors (enterprises, units, and individuals have direct sale behavior with food manufacturers), suppliers (enterprises, units, and individuals can provide food production enterprises with raw materials, equipment, and services), and raw materials (raw materials, subsidiary materials, food additives, food processing aids, and other relevant products used in the food production process).

4.4.1 Basic Requirements for System Construction

The food safety management system, employee's health management system, examination, and checking record system of raw material purchase, food delivery inspection record system, and food recall system should be established and implemented according to the Food Safety Law of the People's Republic of China to ensure food safety.

Raw materials should be nontoxic and harmless and have qualification certificates of raw material suppliers and corresponding products.

The traceability management system should be established by setting up marks, records, and documents for raw materials, semifinished products, and final products.

All records and documents should be completed correctly. Other supporting documents should provide corresponding explanations and collated as appendixes in a record book.

All records and documents should be maintained for 2 years.

4.4.2 Management Requirements for System Construction

1. Personnel management

Personnel management should assign a specific responsibility to every person. The personnel should be appointed to undertake traceability work, and the

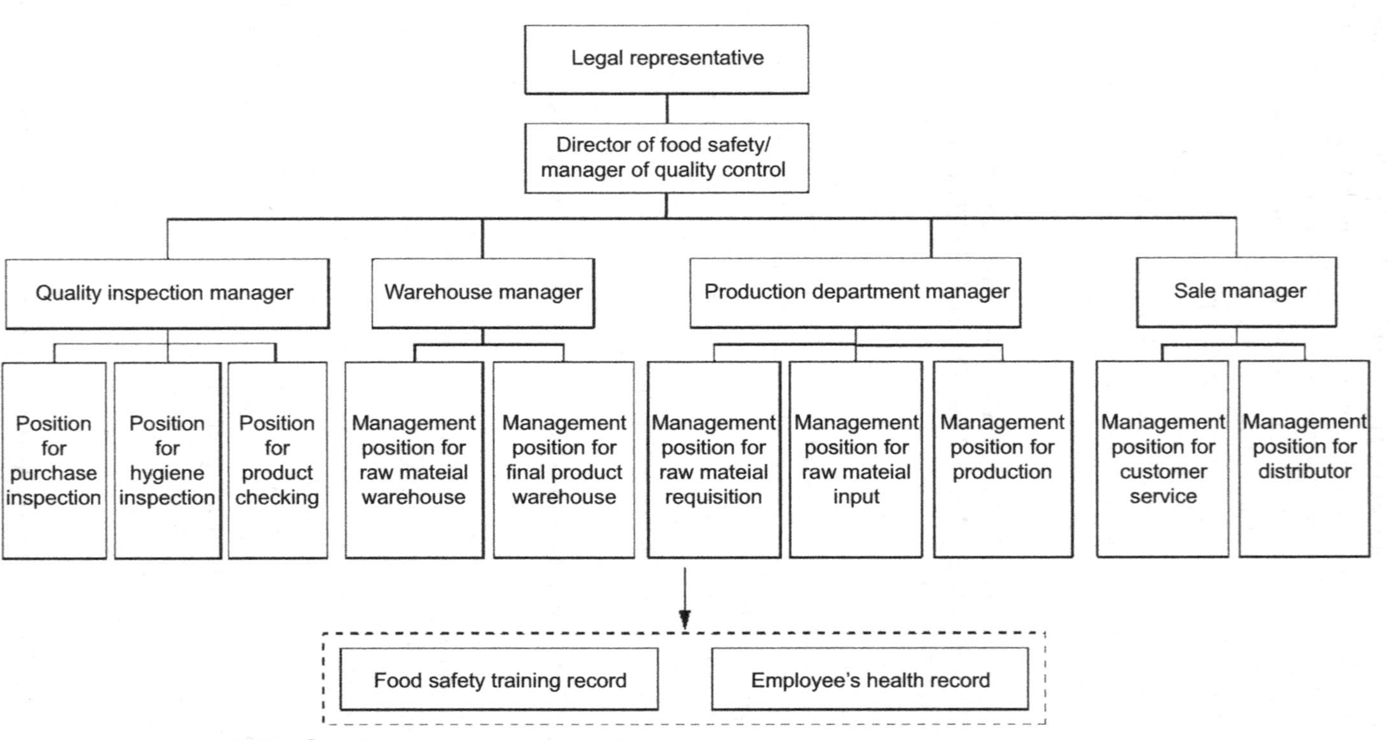

FIGURE 4.1 Organizational structure of the traceability personnel in the food production enterprise.

positions listed in Figure 4.1 should be covered in the enterprise. The organizational structure can be set up according to Figure 4.1.

The person who is in charge of quality inspection, warehouse, production, and sales should review or check the data provided by the staff from his own department and should be responsible for the authenticity and integrity of these data.

2. Management requirements for the examination and acceptance of purchased raw materials

The examination and acceptance of purchased raw materials should be managed according to the requirements shown in Figure 4.2.

3. Assessment of suppliers

The purchasing department, together with relevant departments, will check, evaluate, and collect all qualification certificates of the suppliers. The suppliers that pass checking and assessment will be regarded as the approved suppliers.

After the first batch of goods arriving and passing the inspection, the staff in charge of examination and acceptance of raw materials will record the qualification certificate of the suppliers and establish a supplier list (as shown in A4) and a registration form for the first sale of raw materials (as shown in A3). If the raw materials are the primary agricultural products purchased directly from farmers, a supplier list should be prepared (as shown in A4).

4. Examination and acceptance of raw materials

All raw materials should not be placed in production until it has been determined that these conform to the established quality standards.

A quality inspection certificate should accompany every batch of raw materials that enters the warehouse. The raw materials with the following documents can be regarded as qualified products:

(1) The self-inspection report of raw materials issued by the manufacturers;
(2) The inspection report issued by the qualified inspection institutions;
(3) The sanitary certificate of this batch of imported raw materials issued by the entry-exit inspection and quarantine authorities.

When the supplier is unable to provide the qualification certificate, the enterprise should decide on the inspection items in accordance with the relevant national standards and industry standards to use for the inspection of raw materials and complete the registration form for the inspection of raw materials (as shown in A5) or entrust a qualified inspection agency to perform the corresponding inspection. The raw materials can be used in the production process if it passes the inspection process.

5. Sample preservation of raw materials

The staff in charge of examination and acceptance of raw materials is responsible for completing the registration form for sample preservation of raw materials (as shown in A1).

For important raw materials, the enterprise should preserve a small amount of samples by itself. The amount of preserved samples should be sufficient to meet the established standard examination items of raw materials. The

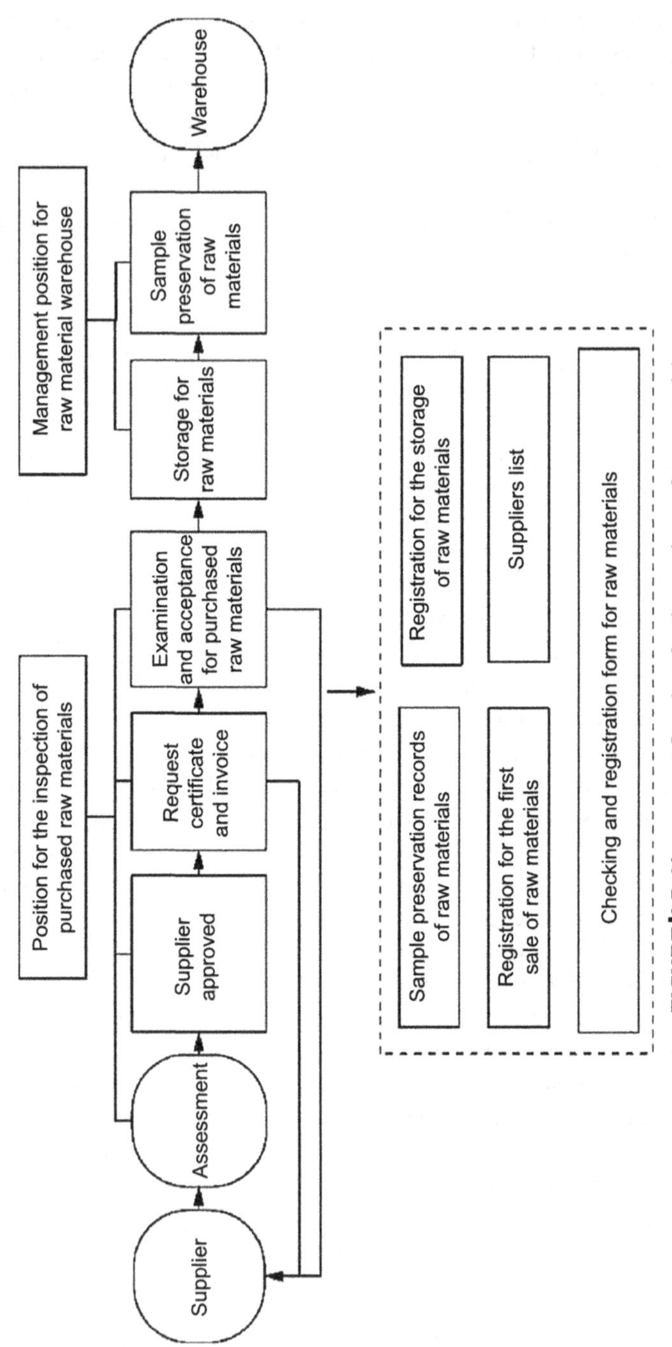

FIGURE 4.2 Management flowchart for the purchase of raw materials.

preservation period of raw materials should not be less than the shelf-life of the final products (except for special raw materials such as fresh agricultural products).

6. Management requirement for storage of raw materials

When the raw materials enter the warehouse, the staff in charge of inspection of purchased raw materials will check the supplier list and the quality examination report of the batch of raw materials and complete the registration form for warehousing of raw materials (as shown in A2). The storage period of these forms should not be less than the shelf-life of final products and should not be less than 2a years. Raw materials from suppliers that are not in the supplier list or raw materials without a quality examination report for the corresponding batch cannot be placed in the warehouse or included in production. Raw materials entering the warehouse should be recorded according to different batches and different purchasing time and labeled which are reflected on their unique batch number.

7. Management requirements for the production process

Management of the production process is shown in Figure 4.3.

Cleaning and disinfection: The cleaning and disinfection of equipment and the hand disinfection of the operating personnel should be recorded. Hygiene inspection staff should regularly check the situation of disinfection and cleaning and complete the cleaning and disinfection inspection record (A8).

The production process: The production department should complete the raw material requisition form (A9), and the material input management staff should complete the registration account form for the use of raw materials (A6). The material input management staff should complete the material input records (A7) once the raw materials enter the practical production phase. The production management staff should complete all relevant production records during the production process. The production records should include the traceability contents such as product name, production date, batch number and lot number, and so forth. The technological parameters of key processes during production should have complete inspection records of key production processes (A11). The warehouse management staff should complete the warehousing record form of the final products (A12), record the traceability contents such as product name, batch number, and quantity, and establish identification marks. The products should not be sold prior to passing this inspection. If there are any remaining raw materials from the production process, the material input management staff or the production management staff should fill out the material-returning form (as shown in A10), and return the extra raw materials to the warehouse. Then, the raw material warehouse management staff should check the material-returning form and make the identification mark for storage. The production management staff may also keep these extra materials and make corresponding marks and record.

8. Management requirements for product inspection

The product inspection staff is responsible for the sample preservation of final products. The amount of preserved samples should meet the requirements for all product inspection items, and the period of sample preservation should be not less than the shelf-life of the final products. The preservation records

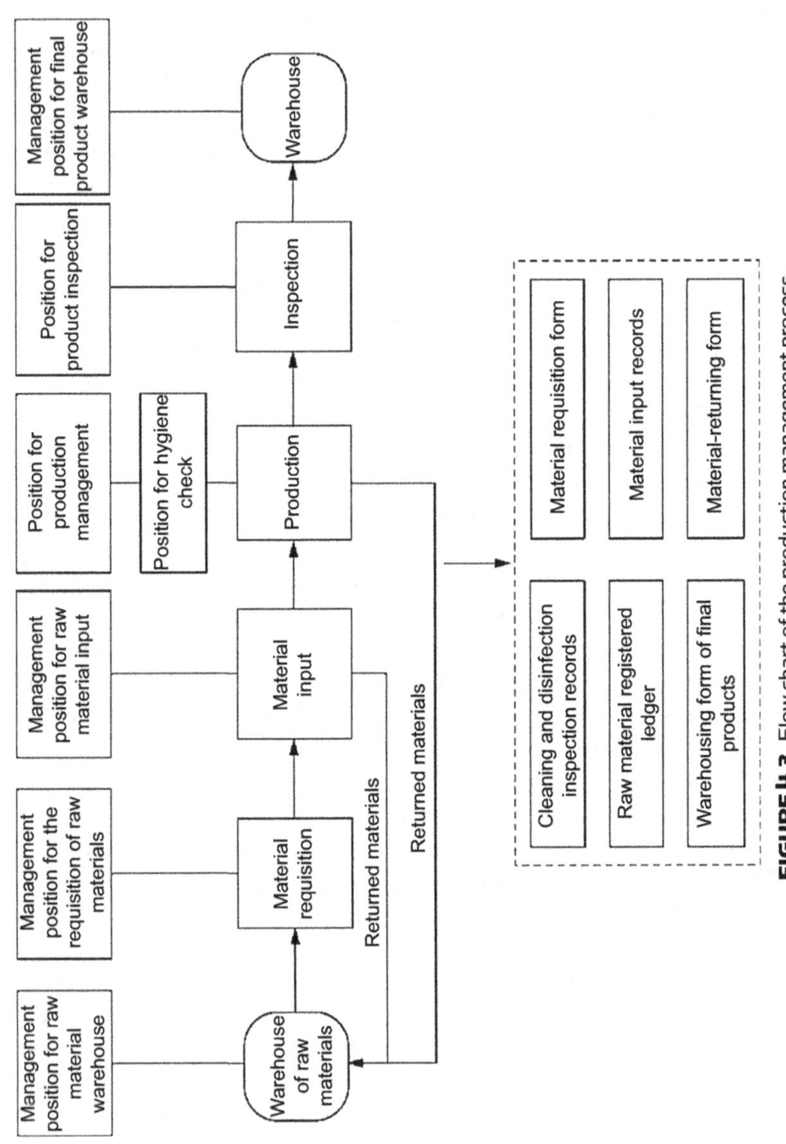

FIGURE 4.3 Flow chart of the production management process.

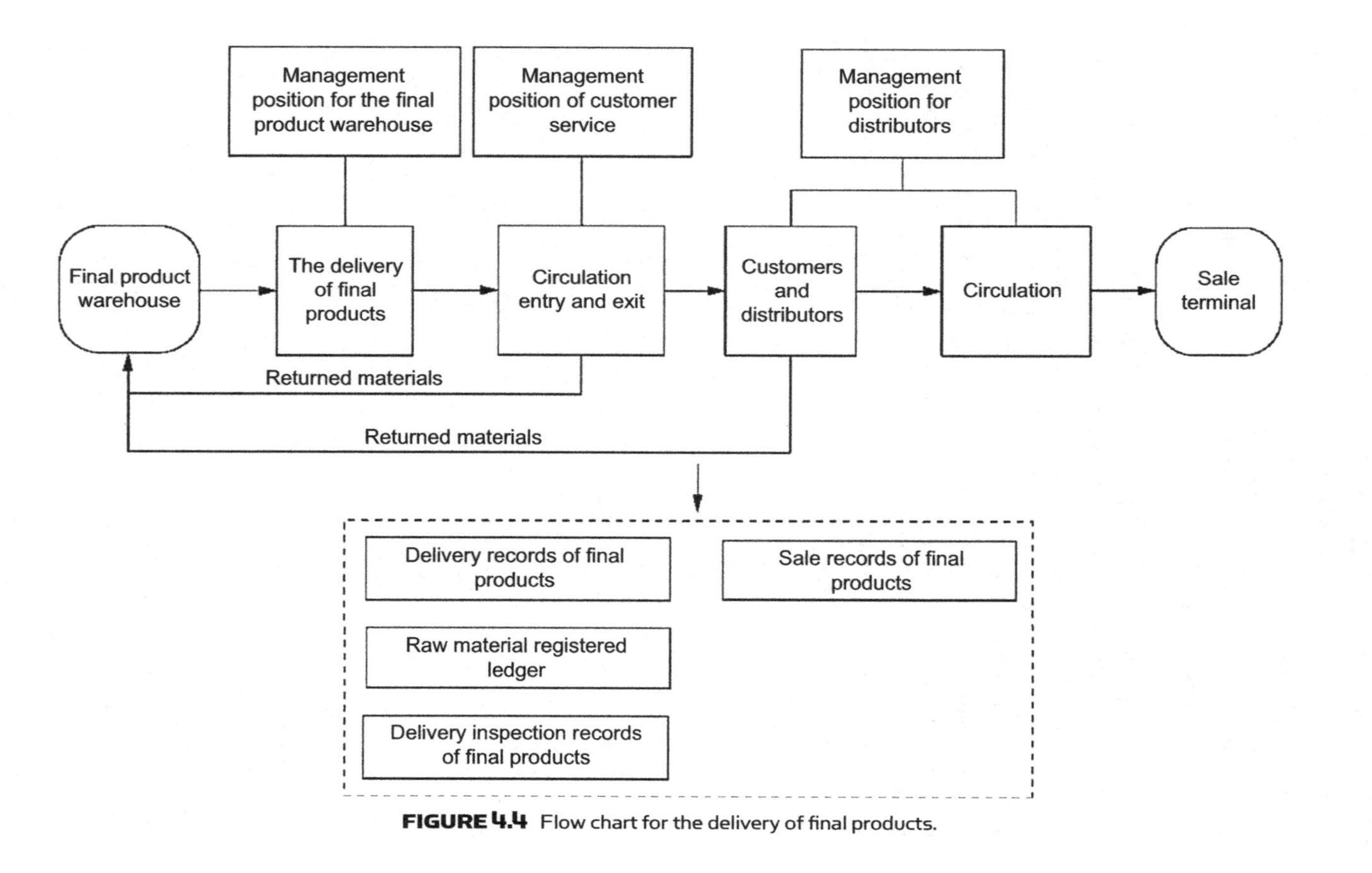

FIGURE 4.4 Flow chart for the delivery of final products.

of final products (as shown in A16) should be completed. The product inspection staff should conduct delivery inspection according to the standards, prepare the original records of the testing items, and complete the "delivery inspection records" (as shown in A15). The person in charge of quality inspection should check these "delivery inspection records."

9. Management requirements for the delivery of final products

The management process for the delivery of final products is shown in Figure 4.4.

The final products that have passed inspection can be sold. The warehouse management staff of final products should complete the delivery records (as shown in A13) for the delivery of final products. The sales records (as shown in A14) should be completed during the circulation process of the final products.

10. Traceability of final products

There should be traceable identification tags on the small package of final products. The identification tags should be able to distinguish the final products from different batches. RFID tags can be used as the trace identifier for internal circulation and the transportation package of the final products (Appendix B). The two-dimensional bar code can also be used as a traceable identification tag (Appendix C).

Appendix A

Traceability Management Forms

Enterprises can use the following forms as records and can also design forms in their own styles, but the forms should include the following information.

A.1 REGISTRATION FORM FOR THE SAMPLE PRESERVATION OF RAW MATERIALS

Registration form number:			
Product name		Bar code	
Lot number		Production date	
Batch number		Supplier name	
Specification		Preserved sample amount	
Preservation time		Registrant	
Preservation department		Sampling person	
Shelf-life		Storage requirement	
Type: □ Raw materials □ Subsidiary materials □ Food additives □ Processing aid agents □ Packaging materials □ Container □ Detergent □ Disinfectant			
Note:			

A.2 REGISTRATION FOR WAREHOUSING OF RAW MATERIALS

Warehousing form number:			
Product name		Supplier	
Serial number of raw materials		Product classification	
Batch number		Warehouse number	
Production place		Ingredient	
Brand		Specification	
Name of the warehousing worker		Warehousing time	
Quantity		Units	
Shelf-life		Production time	
Number of the checking form		Inspector	
Relevant certificates:			
Description of raw materials:			

A.3 REGISTRATION FORM FOR THE FIRST SALE OF RAW MATERIALS

Raw material name		Supplier name	
Supplier code certificate		Supplier type □ Producer □ Distributor □ Wholesaler	
Brand		Carrier	
Bar code/batch number		Raw material type	
Specification		Ingredient	
Manufacturing place		Usage	
Shelf-life		Transportation conditions	
Storage condition		Inspector	
Number of the quality inspection form:			
Relevant certificates:			
Note:			

A.4 SUPPLIER LIST

Supplier list form number:						
Name of the supplier	Supplier type	Address	Contact person	Contact telephone	Name of the raw material	Note

A.5 REGISTRATION FORM FOR THE INSPECTION OF RAW MATERIALS

Inspection form number:		Inspector:	
Raw material name		Raw material type	□ Raw materials □ Subsidiary materials □ Packaging materials □ Food additives
Name of the supplier		Registrant	
Batch number		Registration date	
Quality inspection form number		Conclusion	Qualified/unqualified
Inspection details:			
Testing index	Testing standard	Testing results	Remark

A.6 REGISTRATION ACCOUNT FOR THE CONSUMPTION OF RAW MATERIALS

Consumption form number for raw materials:			
Name of raw materials		Batch number of raw materials	
Name of the final product/semifinished product		Batch number of the final product/semifinished product	
Consumption quantity		User	
Unit		Use time	
Warehouse number		Warehouse keeper	
Note:			

A.7 MATERIAL INPUT RECORD

Material input record form number:			
Product name		Batch number	
Number of the product			
Registrant		Production date	
Usage of raw materials:			
Usage of additives:			
Remark:			

A.8 CLEANING AND DISINFECTION INSPECTION RECORD

Cleaning and disinfection record form number:		Inspector:	
Name of the manufacturer		Type of manufacturer	
Inspection time		Registrant	
Testing place		Registration time	
Inspection details:			
Sanitation index	Testing standard	Testing results	Remark

A.9 MATERIAL REQUISITION

Material requisition form number:			
Name of the raw material		Batch number	
Material quantity		Unit	
Material requisition time		Material requisition person	
Warehouse number		Warehouse keeper	
Remark:			

A.10 MATERIAL-RETURNING FORM

Material-returning form number:			
Name of the raw material		Batch number	
Quantity of returned material		Unit	
Person of material returning		Time of material returning	
Warehouse number		Warehouse keeper	
Remark:			

A.11 INSPECTION RECORDS OF THE KEY PRODUCTION PROCESS

Key production process record form number:				
Product name		Batch number		
Production date		Batch quantity		
Registrant		Registration time		
Order number	Name of key control points	Time of key control point inspection	Inspection results	Remark
1				
2				
3				
4				
5				
6				
7				
8				
9				
10				

A.12 WAREHOUSING OF FINAL PRODUCTS

Warehousing form number:			
Name of the product		Supplier	
Bar code/batch number		Category	
Batch number of the product		Warehouse number	
Place of production		Ingredient	
Brand		Specification	
Warehouse keeper		Warehousing time	
Quantity		Unit	
Operating order of production		Person in charge of production	
Shelf-life		Time of production	
Product checking form number		Inspector	
Relevant certificates:			
Description of raw materials:			

A.13 DELIVERING RECORDS

Delivering record form number:			
Number of product delivering:			
Name of the product		Bar code	
Internal number of products		Warehouse number	
Batch number of the product		Quantity	
Unit:			
Place of production		Specification	
Brand		Time of warehousing	
Warehouse keeper		Time of production	
Shelf-life		Time of delivering	
Number of the inspection form		Inspector	
Relevant certificates:			
Description of raw materials:			

A.14 SALES RECORDS

Sales record form number:			
Number of delivery		Name of the product	
Bar code/internal number		Category	
Batch number		Warehouse number	
Place of production		Ingredient	
Brand		Specification	
Shelf-life		Time of production	
Time of delivery		Sale regions	
Carrier unit		Carrier	
Number of the product checking form		Inspector	
Relevant certificates:			
Description of raw materials:			

A.15 DELIVERY INSPECTION RECORDS

Delivery inspection record form number:		Inspector:	
Name		Type	
Batch number		Registrant	
Number of the quality testing form		Registration time	
Conclusion		Qualified/unqualified	
Testing details:			
Testing index	Testing standard	Testing results	Remark

A.16 SAMPLE PRESERVATION RECORDS OF FINAL PRODUCTS

Sample preservation record form number:			
Name of the product		Bar code	
Batch number of the product		Production date	
Quantity of preserved samples		Preservation period	
Date of sample preservation		Person of sample preservation	
Storage period		Requirements for storage	
Remark:			

Appendix B

Examples of RFID Traceability Identification Tags

B.1 DEFINITION

The following definitions are used for this appendix.

B.1.1 Radio Frequency Identification

Radio frequency identification (RFID) refers to a noncontact automatic recognition technology using transmission characteristics of radiofrequency signals and space coupling (inductive coupling or electromagnetic coupling) to identify the product.

B.1.2 RFID Tag

It is a noncontact information carrier that can identify the product by radiofrequency.

B.1.3 RFID Reader

An electronic equipment for reading or/and writing data in the RFID tag.

B.2 EXAMPLES OF FOOD CODES

B.2.1 Code Structure

For example:

1234-4321-06-0A-09072215626-55

Among these:
1234, Manufacturer identification code, expressing 65535 food production enterprises;
4321, Product code, expressing the product (or the combination) from 65535 specifications;
06, Product quantity within the packaging;
0A, Number of the production line, expressing 256 production lines of the above products;
0907221526, 15:26, July 22, 2009;
55, Serial number of the product, representing the serial number among 256 products.

B.2.2 The Meaning of the Code

The enterprise produces the product with No. 55 package containing six minimum packaging units in the production line A at 15:26 on July 22, 2009.

Note: The checking code of data transmission is checked by the Cyclical Redundancy Code conformed to Standard ISO18000-6B/6C.

Appendix C

Informative Appendix

Examples of two-dimensional bar code traceability identification.

C.1 EXAMPLES OF THE BAR CODE

C.1.1 Bar Code Structure

For example:

(01) 15401234100003.

Among these:

01, Application identifier, the following 14-digit number is the 14-bit product code;

1540, Production enterprise code;

12, Packaging type code;

3, Checking code;

4100003, Product code.

C.1.2 The Meaning of Code

The enterprise that produces the product with a certain packaging type.

C.1.3 Two-Dimensional Bar Code

Two-dimensional bar code is shown in Figure C.1.

FIGURE C.1 Two-dimensional bar code.

C.2 EXAMPLES OF THE PROCESSING PRODUCT CODE

C.2.1 Code Structure

Structure of the processing product code is shown as:

(01) 96901234100016.

Among these:

01, Application identifier, the following 14-digit number is the 14-bit product code;

9690, Production enterprise code;

12, Packaging type code expressing the product with a certain package type;

3, Checking code;

4100016, Product code indicating the enterprise that produced the product.

C.2.2 The Meaning of the Code

The enterprise produces the processing product with a certain packaging type.

C.2.3 The expression of Bar Code

As shown in Figure C.2.

(01) 96901234100016

FIGURE C.2 Two-dimensional bar code of the processing product.

Appendix D

Standard Catalog of Canned Citrus and Codex Stan 254-2007

Standard Catalog of Canned Citrus and Its Products at Home and Abroad

Serial number	Standard number	Title
1	CODEX STAN 254-2007	Standard for certain canned citrus fruits
2	CODEX STAN 078-1981	Standard for canned fruit cocktail
3	CAC/RCP 2-1969	Code of hygienic practice for canned fruit and vegetable products
4	GB/T 13210-1991	Canned mandarin oranges in syrup
5	QB 1393-1991	Canned orange sac coating suspension
6	GB 11671-2003	Hygienic standards for canned fruits and vegetables
7	GB/T 10786-2006	Analytical methods of canned foods
8	GB/T 4789.26-2003	Microbiological examination of food hygiene—commercial sterility testing of canned foods
9	GB/T 5009.69-2008	Analytical methods of hygienic standards for the inner wall of food cans with epoxy phenol coating
10	GB/T 20938-2007	GMP for canned food enterprises
11	NY/T 1047-2006	Green foods, canned fruits, and vegetables
12	SN/T 2100-2008	Rapid determination of commercial sterility in canned foods
13	SN/T 0856-2011	Detection methods of tin in imported and exported canned foods
14	SN/T 0400.1-2005	Rules for the inspection of imported and exported canned foods—Part 1: General principles
15	SN/T 0400.2-2005	Rules for the inspection of imported and exported canned foods—Part 2: Raw and auxiliary materials

Continued...

—Continued

Serial number	Standard number	Title
16	SN/T 0400.3-2005	Rules for the inspection of imported and exported canned foods—Part 3: Processing sanitation
17	SN/T 0400.4-2005	Rules for the inspection of imported and exported canned foods—Part 4: Container
18	SN/T 0400.5-2005	Rules for the inspection of imported and exported canned foods—Part 5: Filling and sealing
19	SN/T 0400.6-2005	Rules for the inspection of imported and exported canned foods—Part 6: Thermal processing
20	SN/T 0400.7-2005	Rules for the inspection of imported and exported canned foods—Part 7: Canned products
21	SN/T 0400.8-2005	Rules for the inspection of imported and exported canned foods—Part 8: Package
22	SN/T 0400.9-2005	Rules for the inspection of imported and exported canned foods—Part 9: Labels

Note: Please refer to the CAC Website if you want to obtain more information regarding the CAC standard text http://www.codexalimentarius.org.
For more information on the national standard text, please refer to the national literature sharing service platform http://www.cssn.net.cn.

D.1 CODEX STANDARD FOR CERTAIN CANNED CITRUS FRUITS (CODEX STAN 254-2007)[1]

D1.1 Scope

This Standard applies to certain canned citrus fruits, as defined in Section D.1.2 below, and offered for direct consumption, including for catering purposes or for repacking if required. It does not apply to the product when indicated as being intended for further processing.

D.1.2 Description

D.1.2.1 PRODUCT DEFINITION

Canned citrus fruit is the product:

1. prepared from washed, sound, and mature ripe grapefruit (*Citrus paradise* Macfadyen), mandarin oranges (*Citrus reticulate* Blanco, including all the suitable commercial varieties for canning), sweet orange varieties (*Citrus sinensis* (L.), Osbeck, including all the suitable commercial varieties for canning), or pummelo (*Citrus maxima* Merr. or *Citrus grandis* (L.));
2. packed with water or other suitable liquid packing medium, sugars as defined in the Codex Standard for Sugars (Codex Stan 212-1999), honey as defined

[1]This Standard supersedes individual standards for canned grapefruit (Codex Stan 15-1981) and canned mandarin oranges (Codex Stan 68-1981).

in the Codex Standard for Honey (Codex Stan 12-1981), suitable spices or flavoring ingredients appropriate to the product;

3. processed by heat, in an appropriate manner, before or after being hermetically sealed in a container, so as to prevent spoilage. Before processing, the fruit shall have been properly washed and peeled and the membrane, seeds and core and fiber strands originating from albedo or core, shall have been substantially removed from the sections.

D.1.2.2 COLOR TYPES (CANNED GRAPEFRUIT OR CANNED PUMMELO ONLY)

1. White: produced from white-fleshed grapefruit or pummelo
2. Pink: produced from pink or red-fleshed grapefruit or pummelo
3. Pale yellow: produced from pale yellow fleshed pummelo

D.1.2.3 STYLES

D.1.2.3.1 Definitions of Styles

Product	Whole[a]	Broken	Twin	Pieces
Canned grapefruit Canned sweet orange	Not less than 75% of original segment	Less than 75% of original segment		
Canned pummelo	Not less than 50% of original segment	Less than 50% of original segment		Large enough to remain on a screen having 8-mm^2 openings formed by a wire of 2-mm diameter
Canned mandarin orange	Not less than 75% of original segment	Not less than 50% of original segment but large enough to remain on a screen having 8-mm^2 openings formed by wire of 2-mm diameter	See definition for whole except two or three segments joined together, which have not been separated during processing	

[a]*A segment that is split in one place only and is not prone to disintegrate shall be considered whole, but parts of a segment joined by a "thread," or by membrane only shall not be considered "whole."*

D.1.2.3.2 Other Styles (Canned Grapefruit, Mandarin Oranges, Sweet Orange Varieties, and Pummelos)

Any other presentation of the product should be permitted provided that the product:

1. is sufficiently distinctive from other forms of presentation laid down in the Standard;
2. meets all relevant requirements of the Standard, including requirements relating to limitations on defects, drained weight, and any other requirements

that are applicable to that style that most closely resembles the style or styles intended to be provided for under this provision;
3. is adequately described on the label to avoid confusing or misleading the consumer.

D.1.2.4 SIZES IN WHOLE SEGMENT STYLE (CANNED MANDARIN ORANGES ONLY)

D.1.2.4.1 Designation in Accordance with Size

Canned mandarin oranges in whole segment style may be designated according to size in the following manner:

1. Uniform single size.
   a. "Large": 20 or less whole segments per 100 g of drained fruit.
   b. "Medium": 21 to 35 whole segments per 100 g of drained fruit.
   c. "Small": 36 or more whole segments per 100 g of drained fruit.
   d. Single sizes shall also meet the uniformity requirements of Section D.1.3.2.5.
2. Mixed sizes.
   A mixture of two or more single sizes.

D.1.3 Essential Composition and Quality Factors

D.1.3.1 COMPOSITION

D.1.3.1.1 Basic Ingredients

Citrus fruit as defined in Section D.1.2 and liquid packing media appropriate to the product.

D.1.3.1.2 Packing Media

In accordance with the Codex Guidelines on Packing Media for Canned Fruits (CAC/GL 51-2003).

D.1.3.1.3 Other Permitted Ingredients (Canned Grapefruit Only)

- Spices.

D.1.3.2 QUALITY CRITERIA

The product shall have color, flavor, odor, and texture characteristic of the product.

D.1.3.2.1 Color

The color shall be typical of fruit that has been properly prepared and properly processed. The liquid packing medium shall be reasonably clear except when it contains fruit juice in compliance with the Codex General Standard for Fruit Juices and Nectars (CODEX STAN 247-2005).

D.1.3.2.2 Flavor

Canned grapefruit, canned mandarin oranges, canned sweet orange varieties, and canned pummelo shall have a normal flavor and odor free from flavors or odors foreign to the product. Canned grapefruit with special ingredients shall have a flavor characteristic of that imparted by the grapefruit and the other substances used.

D.1.3.2.3 Texture

The texture shall be reasonably firm and characteristic for the canned product and reasonably free from dry cells or fibrous portions affecting the appearance or edibility of the product. Whole segments shall be practically free from signs of disintegration.

D.1.3.2.4 Wholeness

For canned grapefruit, canned pummelo, or canned sweet orange varieties only: In the style of whole sections or segments, not less than 50% by weight of drained fruit shall be in whole segments.

D.1.3.2.5 Uniformity of Size

For canned mandarin oranges (whole segment style—single sizes only): In the 95%, by count, of units (excluding broken segments) that are most uniform in size, the weight of the largest unit shall be no more than twice the weight of the smallest unit.

D.1.3.2.6 Defects and Allowances

1. For canned grapefruit, canned sweet orange varieties, and canned pummelo: The finished product shall be prepared from such materials and under such practices that it shall be reasonably free from extraneous fruit matter such as peel or core or albedo and shall not contain excessive defects whether specifically mentioned in this Standard or not. Certain common defects should not be present in amounts greater than the following limitations:
   a. The total surface covered by membrane shall not exceed 20 cm^2 per 500 g of total contents.
   b. Developed seeds shall not exceed 4 per each 500 g of total contents. A developed seed is defined as a seed that measures more than 9 mm in any dimension.
   c. Not more than 15% by weight of the drained fruit may be blemished units. A blemished unit is a fruit section or any portion thereof that is damaged by lye peeling, by discoloration, or by any other visible injury.

2. For canned mandarin oranges:
   The product shall be substantially free from defects within the limits set forth as follows:

Defect	Maximum limit in the drained fruit
• Broken segments (as defined in 2.3.1) (whole segment style)	10% m/m
• Broken segments (as defined in 2.3.1) (twin segment style)	15% m/m
• Membrane (aggregate area)	7 cm^2/100 g (based on sample average)
• Fiber strands (aggregate length)	5 cm/100 g (based on sample average)
• Seeds (that measure more than 4 mm in any dimension)	1/100 g (based on sample average)

D.1.3.3 Classification of "Defectives"

For canned grapefruit, canned mandarin oranges, canned sweet orange varieties, and canned pummelo: A container that fails to meet one or more of the applicable quality requirements, as set out in Sections D.1.2.4 and D.1.3.2 (except those based on sample averages), should be considered as "defective."

D.1.3.4 Lot Acceptance

1. For canned grapefruit, canned mandarin oranges, canned sweet orange varieties, and canned pummelo:
   A lot should be considered as meeting the applicable quality requirements referred to in Sections D.1.2.4 and D.1.3.2 when the number of "defectives," as defined in Section D.1.3.3, does not exceed the acceptance number (c) of the appropriate sampling plan with an AQL of 6.5.
2. For canned mandarin oranges:
   The lot must comply with requirements of Section D.1.3.2.6(2) that are based on sample average.

D.1.4 Food Additives

D.1.4.1 ACIDITY REGULATORS

All Acidity Regulators in Table 3 and in Food Category 04.1.2.4 of the Codex General Standard for Food Additives (CODEX STAN 192-1995).

For mandarin oranges, sweet orange varieties, and pummelos: at the maximum levels established by the GSFA.

INS No.	Name of the food additive	Maximum level
330	Citric acid	GMP (grapefruits)

D.1.4.2 FIRMING AGENTS—FOR ALL CITRUS FRUITS COVERED BY THE STANDARD

INS No.	Name of the food additive	Maximum level
327	Calcium lactate	GMP
509	Calcium chloride	

D.1.5 Contaminants

D.1.5.1 PESTICIDE RESIDUES

The products covered by the provisions of this Standard shall comply with those maximum pesticide residue limits established by the Codex Alimentarius Commission for these products.

D.1.5.2 OTHER CONTAMINANTS

The products covered by the provisions of this Standard shall comply with those maximum levels for contaminants established by the Codex Alimentarius Commission for these products.

D.1.6 Hygiene

1. It is recommended that the products covered by the provisions of this Standard be prepared and handled in accordance with the appropriate sections of the Recommended International Code of Practice-General Principles of Food Hygiene (CAC/RCP 1-1969), Recommended International Code of Hygienic Practice for Low-Acid and Acidified Low-Acid Canned Foods (CAC/RCP 23-1979), and other relevant Codex texts such as codes of hygienic practice and codes of practice.
2. The products should comply with any microbiological criteria established in accordance with the Principles for the Establishment and Application of Microbiological Criteria for Foods (CAC/GL 21-1997).[2]

D.1.7 Weights and Measures

D.1.7.1 FILL OF CONTAINER

D.1.7.1.1 Minimum Fill

The container should be well filled with the product (including packing medium), which should occupy not less than 90% (minus any necessary head space according to good manufacturing practices) of the water capacity of the container. The water capacity of the container is the volume of distilled water at 20 °C that the sealed container will hold when completely filled.

---

[2]For products that are rendered commercially sterile in accordance with the Recommended International Code of Hygienic Practice for Low-Acid and Acidified Low-Acid Canned Foods (CAC/RCP 23-1979), microbiological criteria are not recommended as they do not offer benefit in providing the consumer with a food that is safe and suitable for consumption.

D.1.7.1.2 Classification of "Defectives"

A container that fails to meet the requirement for minimum fill of Section D.1.7.1.1 should be considered as "defective."

D.1.7.1.3 Lot Acceptance

A lot should be considered as meeting the requirement of Section D.1.7.1.1 when the number of "defectives," as defined in Section D.1.7.1.2, does not exceed the acceptance number (c) of the appropriate sampling plan with an AQL of 6.5.

D.1.7.1.4 Minimum Drained Weight

The minimum drained weight shall be as follows

1. For canned grapefruit, canned sweet orange varieties, and pummelos: The drained weight of the product shall be not less than 50%, calculated on the basis of the weight of distilled water at 20 °C that the sealed container will hold when completely filled.
2. For canned pummelo: The drained weight of the product shall be not less than 40%, calculated on the basis of the weight of distilled water at 20 °C that the sealed container will hold when completely filled.
3. For canned mandarin oranges: The drained weight of the product shall be not less than 56%, calculated on the basis of the weight of distilled water at 20 °C that the sealed container will hold when completely filled.[3]

D.1.7.1.4.1 Lot Acceptance. The requirements for minimum drained weight should be deemed to be complied with when the average drained weight of all containers examined is not less than the minimum required, provided that there is no unreasonable shortage in individual containers.

D.1.8 Labeling

D.1.8.1 PRODUCT

The products covered by the provisions of this Standard shall be labeled in accordance with the Codex General Standard for the Labeling of Prepackaged Foods (Codex Stan 1-1985). In addition, the following specific provisions apply:

---

[3]For nonmetallic rigid containers such as glass jars, the basis for the determination should be calculated on the weight of distilled water at 20 °C that the sealed container will hold when completely filled less 20 mL.

D.1.8.2 NAME OF THE PRODUCT

The name of the product shall be "grapefruit," "mandarin oranges," "pummelo," or "oranges," as defined in Section D.1.2.1.

For canned grapefruit, sweet orange varieties, and canned pummelo:

1. The style shall be included as part of the name or in close proximity to the name of the product as in Section D.1.2.3.1.
2. The packing medium shall be included as part of the name or in close proximity to the name of the product as in Section D.1.3.1.2.
3. The color for grapefruit or pummelo if "pink," the color type "pink" shall be included as part of the name or in close proximity to the name of the product.

If an added ingredient, as defined in Section D.1.3.1.3, alters the flavor characteristic of the product, the name of the food shall be accompanied by the term "Flavored with X" or "X Flavored" as appropriate.

For canned mandarin oranges:

1. The style, as appropriate, shall be declared as a part of the name or in close proximity to the name of the product, as follows:
   a. Whole segments. A size classification for whole segments style may be stated on the label if the pack complies with the appropriate requirements of Section D.1.2.4.1 of this Standard. In addition, the number of units present in the container may be shown by a range of count, e.g., "(number) to (number) whole segments."
   b. Broken segments.
2. In the case of sizes, size designation may be declared in close proximity to the style designation, e.g., "mixed sized whole segments."
3. The packing medium shall be declared as part of the name or in close proximity to the name as in Section D.1.3.1.2.

Other styles:

If the product is produced in accordance with the other styles provision (Section D.1.2.3.2), the label should contain in close proximity to the name of the product such additional words or phrases that will avoid misleading or confusing the consumer.

D.1.8.3 LABELING OF NONRETAIL CONTAINERS

Information for nonretail containers shall be given either on the container or in accompanying documents, except that the name of the product, lot identification, and the name and address of the manufacturer, packer, distributor or importer, as well as storage instructions, shall appear on the container. However, lot identification, and the name and address of the manufacturer, packer, distributor ,or importer may be replaced by an identification mark, provided that such a mark is clearly identifiable with the accompanying documents.

D.1.9 Methods of Analysis and Sampling

Provision	Method	Principle	Type
Calcium	NMKL 153:1996	Atomic absorption spectrophotometry	II
	AOAC 968.31 (Codex general method for processed fruits and vegetables)	Complex titrimetry	III
Drained weight	AOAC 968.30 (Codex general method for processed fruits and vegetables)	Sieving gravimetry	I
Fill of containers	CAC/RM 46-1972 (Codex general method for processed fruits and vegetables)	Weighing	I
Solids (soluble)	AOAC 932.12 ISO 2173:1978 (Codex general method for processed fruits and vegetables)	Refractometry	I

D.2 DETERMINATION OF WATER CAPACITY OF CONTAINERS (CAC/RM 46-1972)[4]

D.2.1 Scope

This method applies to glass containers.[5]

D.2.2 Definition

The water capacity of a container is the volume of distilled water at 20 °C that the sealed container will hold when completely filled.

D.2.3 Procedure

1. Select a container that is undamaged in all respects.
2. Wash, dry, and weigh the empty container.
3. Fill the container with distilled water at 20 °C to the level of the top thereof, and weigh the container thus filled.

D.2.4 Calculation and Expression of Results

Subtract the weight found in 3.2 from the weight found in 3.3. The difference shall be considered to be the weight of water required to fill the container. Results are expressed as milliliters of water.

---

[4]As amended by the Committee on Methods of Analysis and Sampling, ALINORM 03/23, Appendix VI-H.
[5]For determination of water capacity in metal containers the reference method is ISO 90.1:1986.

References

Adams, B., Kirk, W., 1991. Process for Enzyme Peeling of Fresh Citrus Fruit. US, 5000967A2. March 19.

Braddock, R.J., 1999. Handbook of Citrus By Products and Processing Technology. A Wiley-Intersicence Publication. John Wiley & Sons, Inc.

Buruusu, A., Uiriamu, K., 1991. Enzymatic Peeling Method for Fresh Citrus Fruit. JP, 3015372. January 23.

Cajustea, J.F., Gonzâlez-Candelasa, L., Veyrata, A., et al., 2010. Epicuticular wax content and morphology as related to ethylene and storage performance of 'Navelate' orange fruit. Postharvest Biology and Technology 55 (1), 29–35.

Carlioz, P., 2005. Domestic Tool for Peeling e.g. Citrus Fruit. FR, 2867371. September 16.

Chen, B., 2003. Food Processing Machinery and Equipments. Machinery Industry Press, Beijing.

Chen, X.F., 2009. Manual for Export Citrus Safety and Quality Control. China Agricultural Press, Beijing.

College of Wuxi Light Industry, College of Tianjin Light Industry, 1981. Food Factory Machinery and Equipment. China Light Industry Press, Beijing.

College of Wuxi Light Industry, Shanghai Design Institute of Light Industry, 1990. Basis of Food Factory Design. China Light Industry Press, Beijing.

Cui, J.Y., 2004. Food Processing Machinery and Equipments. China Light Industry Press, Beijing.

Deng, X.X., 2001. Import & export of China's citrus fruit and its products: current situations and future prospects. World Agriculture (10), 23–25.

Deng, Y.Y., 2010. Optimization of Fast Freezing Citrus Segments Processing Technologies and its Quality Improvement. Midsouth University, Changsha.

Deng, Y.Y., Shan, Y., Li, G.Y., et al., 2009. Study on removing segment membrane of quick frozen *Satsuma mandarin* by enzyme method. Food and Machinery (6), 33–36 61.

Deng, Y.Y., Shan, Y., Li, G.Y., et al., 2010. Optimization of quick freezing process of citrus petals. Chinese Journal of Food Science 31 (6), 288–291.

Department of Agriculture of Hunan Province, 2001. Review and Prospect of Hunan Citrus over a Century. Hunan Science and Technology Press, Changsha.

Department of Development and Plan, Ministry of Agriculture, 2005. Compilation of Advancing Advantage Produce Area Position Distribution Planning. China Agricultural Press, Beijing.

Editorial Board, 1980. Canned Food Industry Manual (5th). China Light Industry Press, Beijing.

Elliott, R.S., Tinibel, J.C., 1993. Enzyme Infusion Process for Preparing Whole Peeled Citrus Fruit. US, 5200217A2. April 6.

Ge, Y.Q., Chen, Y., Zhang, Z.H., et al., 2005. Review on the development of fruit, vegetable and characteristic resources processing industry of China. Chinese Journal of Food Science 26 (7), 270–274.

Grewal, S., Surinder, P., 2004. Method and Device for Peeling Citrus Fruit. JP, 2004159639. June 10.

He, D.X., Shan, Y., Wu, Y.H., et al., 2011. Simultaneous determination of flavanones, hydroxycinnamic acids and alkaloids in citrus fruits by HPLC-DAD-ESI/MS. Food Chemistry 127 (2), 880–885.

He, J.X., 2008. Present situation existing problems and countermeasures of fruit and vegetable cans industry in China. Food and Machinery 24 (2), 151–155.

He, S.W., 1988. Citrus Manual. Hunan Science and Technology Press, Changsha.

Hu, X.S., Liao, X.J., Chen, F., et al., 2005. The present status and developing trend of fruit and vegetable industry in China. Food and Machinery 21 (3), 4–9.

Ismail, M.A., Huiqin, C., Baldwin, E.A., et al., 2005. Optimizing the use of hydrolytic enzymes to facilitate peeling of citrus fruit. Proceedings of the Florida State Horticultural Society 118, 400–402.

Kirk, W., 1993. Apparatus for Peeling Fresh Fruit. US, 5231921. August 3.

Lan, H.L., Wu, H.J., Sun, Z.G., 2003. Enzymatic peeling of orange. Science and Technology of Food Industry 24 (5), 54–56.

Leading Team in Agricultural Industry, Ministry of Agriculture, 2006. Key Technology in Agricultural Products Processing Selection Report. China Agricultural Press, Beijing.

Li, S.G., Zhang, Q., 2006. Manual for Food Processing Machineries and Equipments. Scientific and Technical Documentation Press, Beijing.

Li, S.H., 2003. Food production in the world and the suggestions for increasing competition of Chinese fruit products in international markets. Journal of China Agricultural University 8 (1), 7–13.

Li, S.Y., 1990. Modern Citrus. Sichuan Science and Technology Press, Chengdu.

Li, Z.J., Tan, X.H., Shan, Y., et al., 2009. Development on the selection of naringinase-producing strain and the enzymatic properties. Food and Machinery 25 (1), 141–145.

Liu, C., Zhou, R.Z., 2000. Manual of Food Additives. Beijing University of Technology Press, Beijing.

Liu, F., Osman, A., Yusof, S., 2004a. Effects of enzyme-aided peeling on the quality of local mandarin (*Citrus reticulata* B.) segments. Journal of Food Processing and Preservation 28 (5), 336–347.

Liu, X.L., Deng, X.X., Wang, X.B., et al., 2003. Investigation report on Brazil citrus. South China Fruits 32 (5), 20–26.

Liu, Z., Shan, Y., Li, X.H., 2004b. Discussion of eliminating white muddy thiny in canned Mandarin. Hunan Agricultural Sciences 3, 56–58.

Liu, Z.G., 1996. Manual of Food Additives. China Light Industry Press, Beijing.

Luo, H.R., Shan, Y., 1997. Practical Fruit Products Processing Technologies. Hunan Science and Technology Press, Changsha.

Maria, T.P., Pedro, L., 1997. Pectic enzymes in fresh fruit processing: optimization of enzymic peeling of oranges. Process Biochemistry 32 (1), 43–49.

Ministry of Health, Standardization Administration, The People's Republic of China, 2003. Methods of Food Hygienic Analysis-Physical and Chemical Section-General Principles (GB/T 5009. 1–2003). Standards Press of China, Beijing. August 11.

Pagan, A., Conde, J., Ibarz, A., et al., 2006. Orange peel degradation and enzyme recovery in the enzymatic peeling process. International Journal of Food Science and Technology 41 (2), 113–120.

Pretel, M.T., Botella, M.A., Amoros, A., et al., 2007. Obtaining fruit segments from a traditional orange variety (*Citrus sinensis* (L.) Osbeck cv. Sangrina) by enzymatic peeling. Europen Food Research and Technology 225 (5–6), 783–788.

Sage, P., 1983. Method for Peeling Citrus Fruit. US, 4394393. July 19.

Shan, Y., 2004a. Citrus industry and its research and development system in Spain. Food and Machinery 20 (1), 48–49.

Shan, Y., 2004b. Introduction of Citrus Processing. China Agricultural Press, Beijing.

Shan, Y., 2008. Current situation, development trend and strategy of China citrus processing industry. Journal of Chinese Institute of Food Science and Technology 8 (1), 1–8.

Shan, Y., 2010a. Current situation and development strategic consideration of the fruits and vegetables processing industry in China. Journal of Chinese Institute of Food Science and Technology 10 (1), 1–9.

Shan, Y., 2010b. Study on current situation of the fruits processing industry in China and its functional properties. Journal of Agricultural Engineering Technologies: Agricultural Products Processing (6), 6–8.

Shan, Y., 2012. Current situation and development trend of the fruits processing industry in China. Journal of Beijing Technology and Business University: Natural Science Edition 30 (3), 1–12.

Shan, Y., He, J.X., 1999. Study on Hunan citrus processing technology. Hunan Agricultural Sciences 3, 36–37.

Shan, Y., He, J.X., Fu, F.H., et al., 2003a. Current situation, strategy and prospects of Hunan citrus industry. Hunan Agricultural Sciences 5, 58–61.

Shan, Y., He, J.X., Li, G.Y., et al., 2003b. Current situation, prospects and strategy of Hunan primary agricultural products processing industry. Research of Agricultural Modernization 24 (4), 303–307.

Shan, Y., Hu, Z.J., Li, G.Y., et al., 2009a. A Method of Fast Freezing Citrus Petals. China, CN200810107582. 8. May 27.

Shan, Y., Li, G.Y., 2007. Progress review of modern biotechnology for citrus industry. Food and Machinery 23 (5), 142–145.

Shan, Y., Li, G.Y., He, J.X., et al., 2007a. An Enzymatic Method of Removing Citrus Petals Membrane. China, CN200610032499. X. April 11.

Shan, Y., Li, G.Y., He, J.X., et al., 2007b. A Microbiological Method of Removing Citrus Petals Membrane. China, CN200710034429. 2. August 8.

Shan, Y., Li, G.Y., He, J.X., et al., 2008a. A Device for Removing Fruits and Vegetables Peels and Segment Membrane. China, CN200720064635. 3. July 23.

Shan, Y., Li, G.Y., He, J.X., et al., 2008b. An Enzymatic Method of Removing Whole Citrus Fruit Petals Membrane. China, CN200710035347. X. January 2.

Shan, Y., Li, G.Y., He, J.X., et al., 2008c. A Biological Method of Removing Citrus Peel. China, CN200710036124. 5. March 26.

Shan, Y., Li, G.Y., Zhang, J.H., et al., 2009b. A method of taking of citrus fruit sac coating using compound enzyme preparation in a self-made device. Chinese Journal of Food Science 3, 141–144.

Shan, Y., Li, G.Y., Zhang, J.H., et al., 2009c. Technical study on peeling of citrus whole fruit by enzymatic methods. Journal of Chinese Institute of Food Science and Technology 9 (1), 107–111.

Shan, Y., Li, W.B., He, J.X., et al., 2004. Current situation and development of Brazil citrus industry. Journal of Hunan Agricultural University: Social Sciences 5 (6), 1–5.

Shan, Y., Luo, H.R., 1997. Study on comprehensive processing utilization of pomelo. Hunan Agricultural Sciences 2, 40–41.

Su, D.L., Li, G.Y., He, J.X., et al., 2012. Progress in application of near infrared spectroscopy to nondestructive detection of big yield fruits' quality in China. Science and Technology of Food Industry 33 (6), 460–464.

Tang, C.H., Peng, Z.Y., 2000. Study on citrus functional components. Sichuan Food and Fermentation 4, 1–7.

The First Editing Room, Standards Press of China, 2001. Compilation of China Food Industry Standard (Canned Food Volume). Standards Press of China, Beijing.

USDA Foreign Agricultural Service, 2012. Citrus: World Markets and Trade[R/OL]. July 26. http://www. fas. usda. gov/data/citrus-world-markets-and-trade.

Wang, Y.R., 2012. Analysis of Chinese cans exporting trade in 2011. Canned Food 94 (2), 46–54.

Wei, R.M., 2006. A country from large production to strong competence in fruits and vegetables industry. Productivity Study 3, 51–52.

Wu, H.J., 2001. Current situation and development prospect of China citrus processing industry. South China Fruits 30 (4), 19–20.

Wu, H.J., 2007. Citrus Processing and Comprehensive Utilization Technologies. Chongqing Press, Chongqing.

Wu, W.H., 1994. Engineering Design for Agricultural Products Processing. China Light Industry Press, Beijing.

Wu, Y.H., Shan, Y., Xiang, D.M., et al., 2009a. Advice on developing citrus industrialization in Huaihua, Hunan. Hunan Agricultural Sciences 2, 102–104 107.

Wu, Y.H., Zhang, Q., Xiao, K., et al., 2009b. The hereditary feature and quantitative analysis of functional active ingredient in fruit of Hunan hybrid orange. Hunan Agricultural Sciences 11, 4–7.

Ye, X.Q., 2005. Citrus Processing and Comprehensive Utilization. China Light Industry Press, Beijing.

Zhang, F., 2011. Statistics and analysis of China fruits products trade in 2010. China Fruits Industry Information 28 (3), 1–9.
Zhang, Q., Shan, Y., Wu, Y.H., 2005. Application of HACCP in the produce of citrus cans for export. Modern Food Science and Technology 21 (1), 104–107.
Zhou, Z.Q., 2012. Nutriology of Citrus Products. Scientific Press, Beijing.
Zhu, B.W., 2005. Practical Food Processing Technologies. Chemical Industry Press, Beijing.

Index

'*Note*: Page numbers followed by "f" indicate figures, "t" indicate tables.'

A

Atmospheric cooling, 35
Automatic bottle-washing machine, 56–59, 57f
  bottle-loading and -unloading devices, 58–59, 59f
  cleaning process, 58
  structure, 57

B

Bacteria-induced swollen cans, 36–37
Bar code
  meaning, 161
  processing product code
    expression, 162, 162f
    meaning, 162
    structure, 162
  structure, 161
  two-dimensional bar code, 161, 161f
Batch high-pressure processing, 42

C

Canned citrus products
  current industrial situation, 5–6, 6f
  GMP control, 123–143
    education and training, 133
    factory buildings and equipment, 125
    factory buildings, structure of, 125–126
    factory environment, 124
    ground and drainage, 127–128
    hand-washing facilities and disinfection pool, 129
    health management, 133–139
    lighting and ventilation facilities, 128–129
    locker room, 129
    management mechanism and personnel, 131–132
    management system, establishment and evaluation of, 142–143
    mechanical equipment, 130–131
    personnel requirements, 132–133
    quality inspection equipment, 131
    quality management, 139–141
    record management, 142
    roof and ceiling, 127
    safety facilities, 126
    tag, 142
    toilet, 130
    walls/doors and windows, 127
    warehouse, 130
    water and steam supply facilities, 128
    workshop isolation, 125, 126t
  HACCP application, 116–123, 117t–120t
  hazard analysis and critical control points, 106–116
  history of development, 3–4
    development stage, 1970s to 1980s, 3
    prosperous stage, 1990, 3–4, 4t–5t
    starting stage, 1959–1969, 3
  industry background, 1–2, 2f
  pesticide residues/contaminants and additives, limits and requirements for, 106, 107–111, 111t
  traceability management system, construction of, 143–150
    system construction, basic requirements for, 143
    system construction, management requirements for, 143–150, 146f, 149f
Canned foods
  background, 8–10
  categories, 10–11
  preservation principles, 11–18
    exhausting effect, 11–12
    heat conductivity effect, 16
    microorganisms, heat tolerance of, 14
    sealing effect, 12
    spoilage microorganisms, 13
    sterilization effect, 13–18
  procedures and techniques
    canning, 24
    cooling, 35
    exhausting, 27
    flowchart, 18, 19f
    key points, 18–26
    peel/pith removing, 20
    precleaning raw citrus fruits, 18
    presealing, 27
    sac coating, removal of, 21
    sealing, 30–32
    sterilization, 32–35

Canned foods (*Continued*)
quality issues and controlling
cans, turbidity and precipitation in, 38
can walls, corrosion of, 37
food discoloration/spoilage, 37–38
swollen cans, 36–37
sterilization techniques, 38–46
electrostatic sterilization, 44
high-pressure processing (HPP), 42–43
induction electronic sterilization, 45
magnetic sterilization, 44
microwave technique, 39
Ohmic sterilization, 39–40
others, 45–46
ozone sterilization technique, 41–42
pulsed electric field (PEF), 43, 44f
pulsed light sterilization technology, 40–41
Canned juice, 11
Canned sauce, 10–11
Canning method, 26
Chain-belt exhausting cabinet, 30
Chitin sterilization, 45
Citrus fruits, 25
Citrus-sorting machine
citrus image-processing sorting machine, 64–66, 64f–65f
citrus optical sorting machine, 63–64, 63f
inherent citrus quality-sorting equipment, 66–72, 67f–69f, 71f
roller-style citrus sorter, 59–60, 60f–61f
three-roller citrus sorter, 60–63, 61f–62f
Cleaning machinery/equipment, 51–59
packaging containers. *See* Packaging containers
raw citrus materials. *See* Raw citrus materials
Clostridium
C. botulinum, 13
C. pasteurianum, 13–14
Codex standard (CODEX STAN 254-2007)
composition
basic ingredients, 166
other permitted ingredients, 166
packing media, 166
contaminants, 169
defectives classification, 168
description
color types, 165
product definition, 164–165
styles, 165–166, 165t
food additives, 168–169, 169t
hygiene, 169
labeling, 170–171
lot acceptance, 168
quality criteria
color, 166
defects and allowances, 167–168
flavor, 167
size uniformity, 167
texture, 167
wholeness, 167
scope, 164
weights and measures, 169–170
Cold water spraying, 58
Containers (CAC/RM 46-1972)
calculation and expression, 172
definition, 172
procedure, 172
scope, 172
Continuous atmospheric sterilization machinery, 34
Continuous high-pressure processing (CHPP), 42–43
Continuous low-temperature sterilization, 34
Continuous sterilization, 34
Controlling enzyme activity, 18
Conventional method, 21
Counterpressure cooling, 35

D

Detergent soaking, 58
Detergent spraying, 58
Discontinuous atmospheric sterilization, 33
Discontinuous high-pressure sterilization, 33

E

Enzymatic hydrolysis conditions, 24
Enzymatic peeling technique, 20
Enzyme selection, 24
Equipment
cleaning, 51–59
packaging containers. *See* Packaging containers
raw citrus materials. *See* Raw citrus materials
exhaust. *See* Exhausting methods
material handling, 48–51
belt-driven conveyors, 48, 48f
bucket elevator, 49–50, 49f–50f
spiral conveyor, 51, 51f
packaging
canned citrus segment products, sealing machinery for, 87–91, 89f–91f
carton-sealing machine, 101–102, 101f
GT786 automatic vacuum juice-filling machine, 86–87, 87f–88f
labeling machine, 96–98, 96f, 98f
packing machine, 98–101, 99f
spinning capper, 91–94, 91f–94f
strapping machine, 102–103, 103f–104f
vacuum-packaging machine, 94–96, 95f

raw citrus materials and semifinished products processing
citrus-sorting machine. *See* Citrus-sorting machine
orange segment-sorting machine, 73, 74f
removing citrus sac coating, pilot equipment for, 73–75, 75f
scraper-style continuous citrus peel-heating machine, 72–73, 72f
sterilization, 79–85
horizontal sterilization equipment, 79–80, 79f
rotary continuous pressure sterilization equipment, 83–85, 83f–86f
rotary sterilization device, 80–83, 81f
typical canned citrus-processing production line, 104, 104f
Ethylenediaminetetraacetic acid (EDTA), 21
Exhausting methods, 29
chain-belt exhaust cabinet, 78
gear disc exhausting cabinet, 76–78, 77f–78f

F

Factory buildings
equipment, 125
structure of, 125–126
Factory environment, 124
Flame sterilization, 13
Food acidity, 29

G

GMP control, 123–143
education and training, 133
factory buildings and equipment, 125
factory buildings, structure of, 125–126
factory environment, 124
ground and drainage, 127–128
hand-washing facilities and disinfection pool, 129
health management, 133–139
lighting and ventilation facilities, 128–129
locker room, 129
management mechanism and personnel, 131–132
management system, establishment and evaluation of, 142–143
mechanical equipment, 130–131
personnel requirements, 132–133
quality inspection equipment, 131
quality management, 139–141
record management, 142
roof and ceiling, 127
safety facilities, 126
tag, 142
toilet, 130
walls/doors and windows, 127
warehouse, 130
water and steam supply facilities, 128
workshop isolation, 125, 126t

H

Hand-washing facilities, disinfection pool, 129
Health management, 133–139
Heating exhaust, 29–30
Heating sterilization, 13
High-frequency sealing method, 31
High-pressure processing (HPP), 42–43
Hot-press sealing method, 31
Hot water spraying, 58
Hydrogen-induced swollen cans, 36
Hydrolock continuous sterilization, 34
Hydrostatic pressure sterilization, 34

I

Induction electronic sterilization, 45
Infrared sterilization, 45

L

Lighting facilities, 128–129
Locker room, 129

M

Machinery
cleaning, 51–59
packaging containers. *See* Packaging containers
raw citrus materials. *See* Raw citrus materials
exhaust. *See* Exhausting methods
material handling, 48–51
belt-driven conveyors, 48, 48f
bucket elevator, 49–50, 49f–50f
spiral conveyor, 51, 51f
packaging
canned citrus segment products, sealing machinery for, 87–91, 89f–91f
carton-sealing machine, 101–102, 101f
GT786 automatic vacuum juice-filling machine, 86–87, 87f–88f
labeling machine, 96–98, 96f, 98f
packing machine, 98–101, 99f
spinning capper, 91–94, 91f–94f
strapping machine, 102–103, 103f–104f
vacuum-packaging machine, 94–96, 95f
raw citrus materials and semifinished products processing
citrus-sorting machine. *See* Citrus-sorting machine
orange segment-sorting machine, 73, 74f

Machinery (*Continued*)
removing citrus sac coating, pilot equipment for, 73–75, 75f
scraper-style continuous citrus peel-heating machine, 72–73, 72f
sterilization, 79–85
horizontal sterilization equipment, 79–80, 79f
rotary continuous pressure sterilization equipment, 83–85, 83f–86f
rotary sterilization device, 80–83, 81f
typical canned citrus-processing production line, 104, 104f
Magnetic sterilization, 44
Management system, establishment and evaluation of, 142–143
Marmalade, 11
Material handling machinery/equipment, 48–51
belt-driven conveyors, 48, 48f
bucket elevator, 49–50, 49f–50f
spiral conveyor, 51, 51f
Mechanical canning, 26
Mechanical equipment, 130–131
Microwave sterilization, 39
technique, 39

N

Nuclear radiation sterilization, 45

O

Ohmic sterilization, 39–40
Optimal sterilization, 33

P

Packaging
canned citrus segment products, sealing machinery for, 87–91, 89f–91f
carton-sealing machine, 101–102, 101f
GT786 automatic vacuum juice-filling machine, 86–87, 87f–88f
labeling machine, 96–98, 96f, 98f
packing machine, 98–101, 99f
spinning capper, 91–94, 91f–94f
strapping machine, 102–103, 103f–104f
vacuum-packaging machine, 94–96, 95f
Packaging containers
automatic bottle-washing machine, 56–59, 57f
bottle-loading and -unloading devices, 58–59, 59f
cleaning process, 58
structure, 57
filled cans, cleaning machine for, 55–56, 56f
three-piece can-cleaning machine, 54, 55f
Phosphate-sodium hydroxide (NaOH), 21
Polysaccharides, 21
Preservation principles, 11–18
exhausting effect, 11–12
heat conductivity effect, 16
microorganisms, heat tolerance of, 14
sealing effect, 12
spoilage microorganisms, 13
sterilization effect, 13–18
Prewashing, 58
Protopectin, 21–22
Pulsed electric field (PEF), 43, 44f
Pulsed light sterilization technology, 40–41
Pulse sealing method, 31–32

Q

Quality inspection equipment, 131
Quality management, 139–141

R

Radiation sterilization, 13
Radio frequency identification (RFID)
defined, 159
food codes, examples of
code structure, 159–160
meaning, 160
reader, 159
tag, 159
Raw citrus materials, 52–54
citrus-sorting machine. *See* Citrus-sorting machine
drum-type fruit-washing machine, 53–54, 53f
DT5A1 fruit-washing machine, 52–53, 53f
GT5A9 fruit-brushing machine, 52, 52f
orange segment-sorting machine, 73, 74f
removing citrus sac coating, pilot equipment for, 73–75, 75f
scraper-style continuous citrus peel-heating machine, 72–73, 72f
Record management, 142

S

Safety facilities, 126
Semifinished products processing
citrus-sorting machine. *See* Citrus-sorting machine
orange segment-sorting machine, 73, 74f
removing citrus sac coating, pilot equipment for, 73–75, 75f
scraper-style continuous citrus peel-heating machine, 72–73, 72f
"Shan Guang 18" sterilization, 34–35

Sterilization
  methods, 33
  techniques, 38–46
    electrostatic sterilization, 44
    high-pressure processing (HPP), 42–43
    induction electronic sterilization, 45
    magnetic sterilization, 44
    microwave technique, 39
    Ohmic sterilization, 39–40
    others, 45–46
    ozone sterilization technique, 41–42
    pulsed electric field (PEF), 43, 44f
    pulsed light sterilization technology, 40–41
Sugar solution cans, 10
Swollen cans, 36–37
Syrup cans, 10

T

Traceability management forms, raw materials registration
  cleaning and disinfection inspection, 154
  consumption, 153
  delivering records, 156
  delivery inspection records, 157
  final products, sample preservation records, 158
  final products warehousing, 156
  first sale, 152
  inspection records, 153, 155
  material input record, 154
  material requisition, 154
  material-returning form, 155
  preservation, 151
  sales records, 157
  supplier list, 153
  warehousing, 152

U

Ultrahigh-pressure sterilization, 35

V

Ventilation facilities, 128–129

W

Warm water spraying, 58
Workshop isolation, 125, 126t

X

X-ray sterilization, 45

Printed in the United States
By Bookmasters